Micro Electro Mechanical System (MEMS) Technology for Multifaceted Applications

Kamaljeet Singh · Ayan Karmakar

Micro Electro Mechanical System (MEMS) Technology for Multifaceted Applications

Springer

Kamaljeet Singh
Semi-Conductor Laboratory (SCL)
Ministry of Electronics and Information
Technology (MeitY), Government of India
Mohali, Punjab, India

Ayan Karmakar
Semi-Conductor Laboratory (SCL)
Ministry of Electronics and Information
Technology (MeitY), Government of India
Mohali, Punjab, India

ISBN 978-981-96-6161-9 ISBN 978-981-96-6162-6 (eBook)
https://doi.org/10.1007/978-981-96-6162-6

This Springer imprint is published by the registered company Springer Nature Singapore Pte Ltd.
The registered company address is: 152 Beach Road, #21-01/04 Gateway East, Singapore 189721, Singapore

Preface

Advent of semiconductor technology using silicon as substrate was introduced in late 1960s after the advent of micromachining. Micro-devices employing semiconductor fabrication techniques are gaining lot of interest due to compact structure and better performance. The ease of fabrication, miniaturization led to the popularity of these devices. These devices can be subset of MEMS but are not having mechanical membrane. The wide acceptance of these devices due to reduction in fabrication cost, size, and power consumption while maintaining better performance led to the widespread integration into various systems. Traditional and bulky mechanical sensors are widely employed in the automobile industry but microfabrication-based pressure sensors, accelerometers have already replaced the bulky sensors. Bulk production, enhanced performance and miniaturization make MEMS technology set apart from the traditional way. Micro-devices are basically offshoot of the MEMS circuits without movable membrane but employing the same fabrication techniques. Recently micromachined acoustic devices such as microphones and microspeakers are being fabricated and tested for applications such as hearing and cellular phone, micropersonal digital assistant (micro-PDA), earphone and microspeaker, etc. Since last decade attempts have been made for the development of acoustic microsensors by integrating a piezoelectric layer of zinc oxide (ZnO). The excellent piezoelectric and dielectric properties of zinc oxide (ZnO) make it an interesting material for investigation because of the potential advantage of its compatibility with microelectronics wet etching techniques, low cost, small volume array configuration. Basic wet chemistry employed in bulk micromachining for microcircuits realization but imposes process constraint as in CMOS fabline it is not allowed to and fro movement due to gate integrity issues resulting from highly mobile ions. The semiconductor technology grew in complexity with respect to number of gate counts, speed and functionality of circuits. This not only makes understanding of silicon in better way but also aggressively moved in the direction of integration of sensors with the CMOS. Mostly, silicon is employed, and various variants such as SOI, epitaxial, and high resistivity wafer are also employed and needs individual process optimization. Realization of micro-device imposes constraint on the process integration and control due to tight fabrication tolerances.

This book details simple methodology for the realization of the practical micro-devices employed in general applications. The book brings out practical concept along with process details associated with the device realization. The micro-device fabrication technique is similar to MEMS, and the application and methodologies are compatible to standard CMOS technologies. This book covers basic of wafer, general micro-devices such as microheater, temperature sensor, humidity sensor, gas sensor, pyro-device along with associated characterization, integration, and assembly aspects. This book provides basic overview and discusses practical aspects with characterization data. This book is targeted for the beginners and researchers providing insight into the practical aspects of this technology.

Mohali, India Kamaljeet Singh
Ayan Karmakar

Acknowledgements

> *In life you will realize there is role for everyone you meet,*
>
> *Some will test you, some will use you, some will love you and some will lead you,*
>
> *But the ones who are truly important are the ones who bring out best in you.*
>
> *They are the rare and amazing people who remind you why it's worth it.*
>
> —*Mahatma Buddha*

This quotation reminds us about the people whom we worked with during our career as we encountered with different personalities and made a point to learn from them. This book is the culmination of the team effort during of the re-establishment and operationalization of 6" fab line at SCL. The effort toward establishment of various fabrication processes, unit step development, process integration, product lot, and characterization resulted in the operationalization of the fab line in record time. The manifold challenges encountered during this process were successfully overcome with the support, guidance and motivation of all senior colleagues and Directors. The sincerity, dedication, integrity, and methodological working truly inspire and motivated us to work toward gaining of cognizance in this area. We are indebted to Secretary-MeitY and Chairman-ISRO, for giving us the opportunity and full access to resources without which it is impossible to finish this work. We are always thankful to the ISRO and MeitY culture for providing the great working environment with competent and inspiring leadership. We would also like to thank our well-wishers, near, and distant, who have kept us in their thoughts and prayers.

Our sincere gratitude to Secretary-MeitY and Chairman-ISRO for all the motivation and constant encouragement. The culture prevalent at ISRO/MeitY motivates and nurtures the quality and work ethics making it at par with the world class organization.

Our sincere thanks goes to various senior officials/scientists for their moral support and contribution toward our development. We thank them from the bottom of our hearts. We are also thankful to various people who all compelled us to give back to the society in various forms. Special thanks are also to the friends and colleagues at SCL-Chandigarh and URSC-Bangaluru, who have directly or indirectly helped us daily to learn this technology from scratch. We learned this subject from researchers and authors who are too many to include in this short space, but we gratefully acknowledge their contributions and providing me new insight into this topic. We express our thankfulness to the Almighty who has given us inner strength in completing this book concurrently with other assignments.

Last but not least, we are most thankful to our families for their boundless love and support, who never failed to believe on us.

Kamaljeet Singh
Ayan Karmakar

Contents

About the Authors

Dr. Kamaljeet Singh joined U R Rao Satellite Centre in November 1999 and worked in the area of RF receiver for satellite as designer. He contributed in various capacity and areas at U R Rao satellite centre notably realization and production of satellite transponders, Head of central fabrication facility responsible for flight cards fabrication. He was head of MEMS Development Division in SCL (2010–2015) responsible for the re-establishment of 6" fab-line responsible for the realization of various sensors. He was Group Director of HMC group from 2020–2023 where he was involved in development of HMCs and MIC/MMIC circuits. Since December 2023 he was heading SCL unit of MeitY. He is Fellow of IETE, Fellow of Punjab Academy of sciences, Life member of ASI and IMAPS and senior member of IEEE. He has published more than 200technical articles in various journals/conferences and has written various technical books. He was awarded by IETE, INAE, and ISRO for his technical contributions.

Ayan Karmakar received his M.Tech. (Electronics and Telecom Engineering) degree from NIT-Durgapur. Currently he is working in SCL, Chandigarh as a Scientist-SF. His research interests include design and development of various passive microwave integrated circuits and antennae using silicon based MIC, RF-MEMS and THz-technology. He has authored one technical book named *Si-RF Technology* (Springer-2019) and co-authored another book titled *Printed Antennas for Future Generation Wireless Communication and Healthcare* (CRC Publication 2023). He has more than 45 publications in refereed journals and conferences. He serves as a reviewer in top tier international journals, like-*IEEE Access*, *RF-MiCAE*, *Springer*, *EJAET*, etc. He is the Fellow of IETE and life member of Indian Science Congress Association and ISSE.

Abbreviations

ADS	Advanced design system
AFM	Atomic force microscopy
Al	Aluminum
Al_2O_3	Alumina/aluminum oxide
AlN	Aluminum nitride
APCVD	Atmospheric pressure CVD process
ASIC	Application specific integrated circuit
Au	Gold
BCB	Benzocyclobutene
BMD	Bulk microdefect
BOE	Buffered oxide etchant
BPF	Bandpass filter
BST	Barium strontium titanate
CAD	Computer-aided design
CATF	Compact antenna test facility
CMOS	Complementary metal oxide semiconductor
CMP	Chemical mechanical polishing
CPW	Coplanar waveguide
CR	Capacitance ratio
Cu	Copper
C-V	Capacitance voltage
CVD	Chemical vapor deposition
CZ	Czochralski method
DGS	Defected ground structure
DI	De-ionized
DMTL	Distributed MEMS transmission line
DRIE	Deep reactive ion etching
DUT	Device under test
EDA	Electronic design automation
EDX	Energy dispersive X-ray
EDXA	Energy dispersive X-ray analysis
EM	Electromagnetic

ESD	Electrostatic discharge
FBAR	Film bulk acoustic resonator
FCC	Federal communication commission
FDTD	Finite difference time domain
FEM	Finite element method
FG-CPW	Finite ground coplanar waveguide
FOM	Figure of merit
FTIR	Fourier transform infrared spectroscopy
FWHM	Full width at half maxima
FZ	Float zone
GaAs	Gallium arsenide
GCPW	Grounded coplanar waveguide
GHz	Gigahertz
GPa	Giga Pascal
GSG	Ground-signal-ground
HF	Hydrofluoric acid
HMICs	Hybrid microwave integrated circuits
HNA	Hydrofluoric acid + Nitric Acid + Acetic Acid
HRS	High resistive silicon
HRSi	High resistive silicon
HTCC	High temperature co-fired ceramic
Hz	Hertz
IC	Integrated circuits
IEEE	Institution of electrical and electronics and engineers
IMD	Intermodulation products
ISM	Industrial and scientific band
K-band	IEEE radar band letter designation for 18 to 26.5 GHz
Ka-band	IEEE radar band letter designation for 26.5 to 40 GHz
Ku-band	IEEE radar band letter designation for 12 to 18 GHz
LCR	Meter by which inductance, capacitance, and resistances are measured
$LiNbO_3$	Lithium niobate
LOR	Lift-off resist
LPCVD	Low pressure chemical vapor deposition
LRS	Low-resistivity silicon
LTCC	Low temperature co-fired ceramic
MAM	Metal-air-metal
MEMS	Micro-electro-mechanical system
MESFET	Metal field effect transistor
MHZ	Megahertz
MIM	Metal-insulator-metal
MMICs	Millimeter wave integrated circuits
MOCVD	Metal oxide chemical vapor deposition
MOEMS	Optical MEMS
MSL	Microstrip line
MST	Microsystem technology

NEMS	Nano-electro-mechanical system
$PbTiO_3$	Lead titanate
PDMS	Polydimethylsiloxane
PECVD	Plasma-enhanced chemical vapor deposition
PMMA	Polymethyl methacrylate
Pt	Platinum
PVDF	Polyvinylidene difluoride
PZT	Lead zirconate titanate
RCA	Radio Corporation of America
RFIC	Radio frequency integrated circuits
RIE	Reactive ion etching
RTD	Resistance temperature detectors
SAW	Surface acoustic wave
SEM	Scanning electron microscope
Si	Silicon
Si_3N_4	Silicon nitride
SiC	Silicon carbide
SiGe	Silicon germanium
SiO_2	Silicon dioxide
SIR	Stepped impedance resonator
SOG	Silicon on glass
SOI	Silicon on insulator
SOQ	Silicon on quartz
SOS	Silicon on sapphire
SPDT	Single pole double throw
T/R	Transmit/receive
T-line	Transmission line
TMAH	Tetra methyl ammonium hydroxide
TSV	Through silicon via
TTD	True time delay
ULPA	Ultra-low particulate air
UV	Ultraviolet
UWB	Ultra-wideband
VHF	Vapor HF
VNA	Vector network analyzer
VSWR	Voltage standing wave ratio
W2W	Wafer to wafer
WiMAX	Worldwide interoperability for microwave access (based on IEEE 802.16)
WLP	Wafer level packaging
XeF_2	Xenon difluoride
XRD	X-ray diffraction
ZMR	Zone metal recrystalline polysilicon
ZnO	Zinc oxide

Symbols

Z_o	Characteristics impedance
Y_o	Characteristics admittance
Y_{ine}	Even-mode input admittance
Y_{ino}	Odd-mode input admittance
Z_{oo}	Odd-mode impedance
Z_{oe}	Even-mode impedance
$\tan\delta$	Loss-tangent
fbw	Fractional bandwidth
W	Width of transmission line
f_c	Cut-off frequency
f_o	Center frequency
f_{res}	Resonant frequency
α	Attenuation constant
c	Coupling coefficient
Γ	Reflection constant
Γ_e	Even-mode reflection constant
Γ_o	Odd-mode reflection constant
$M_{i,j+1}$	Coupling/cross-coupling between resonators
ε	Permittivity
θ_e	Even-mode electrical length
θ_o	Odd-mode electrical length
g_i	Coupling coefficient
Q_E	External quality factor
Q_U	Unloaded quality factor
π	Universal constant (22/7)
σ	Conductivity of the medium
η	Intrinsic impedance
ω	Angular frequency
c	Speed of the light in vacuum
μ	Vacuum permeability
h	Height of the substrate

g_m	Transconductance
f_{max}	Maximum frequency of operation
C_{gs}	Gate-source capacitance
f_T	Transition frequency
T_t	Transit time
τ	Minority carrier life-time
N	Ideality factor
c	Velocity of light
C	Capacitance
h	Plank's constant
e-m	Electromagnetic
k	Boltzmann constant
ω	Angular velocity
σ	Bulk conductivity
Z	Impedance

Chapter 1
Role of Contamination

This chapter details about the sources of different types of contaminations in semiconductor wafers, its role in the reliability of the MEMS devices, and mitigation techniques to mitigate the associated quality issues. It highlights on various cleaning methodologies adopted during realizing of 3D-microstructures for alleviating contaminations.

1.1 Introduction

The reliability issue in semiconductor industry greatly influenced by the contaminations or impurities embedded in situ processes. This specialized branch of electronics field employs multiple process steps, subjected to various pressure and temperature conditions. Photolithography, etching, deposition, metallization, and patterning are the main process categories. Contamination plays a key role in the reliability of IC fabrication technology, as the dimension of the device is also getting shrink day by day with the lower technology nodes, keeping in synergy with gigantic growth in this field [1–3].

Contaminants in the semiconductor wafers can be broadly categorized into three major heads, which are as follows:

(i) Organic on inorganic contamination.
(ii) Ionic contamination.
(iii) Various impurities like metal, chemical, gases, etc.

In case of micro-electro-mechanical system (MEMS) technology, not only the miniaturization of the device plays important role but also the electro-mechanical behavior of the device become utmost important. A single device can fail either

K. Singh and A. Karmakar, *Micro Electro Mechanical System (MEMS) Technology for Multifaceted Applications*, https://doi.org/10.1007/978-981-96-6162-6_1

electrically or mechanically due to embedded impurity involved in a MEMS sensor. In recent times, a standalone MEMS sensor is not at prime choice, rather than its complete solution with readout circuitry, especially developed by CMOS process [4–6].

Residual stress, structural stability, surface roughness, and unique feature of free-hanging structures are some of the unique characterization of MEMS devices as compared to its counterpart CMOS. Apart from these, multiple wafer bonding capability, deposition of various noble and refractory metals, featuring ultra-thick metallization, etc. are also some special features of MEMS/micromachining process [7–9].

In microfabrication process, high failure-rate can be due to engagement of various contaminants, like Na^+, metals, substrate particles on the surface of wafers, etc. In situ high temperature processing steps (deposition, growth, annealing, wafer bonding, etc.) can diffuse these impurities into various device layers. For example, photo-resist particles can be detrimental in device performance and can lead to shorting conditions. The main detrimental effects of such impurities on contaminants are electrical defect, substrate leakage, variation in resistivity, dielectric charging phenomena, etc. Thus, it is directly linked with the yield of device. In micromachining process, the wafer bonding mechanisms (anodic, eutectic, or fusion, mainly) need ultra-clean surface of wafers that means almost ideally free of impurities. Several experimental results demonstrate that the throughput of the bonding process strongly depends upon the cleanliness of the bonding interfaces [10, 11].

The ultimate goal of MEMS technology is the amalgamation of several technological fields, like electrical, mechanical, and chemical physical or optical and obviously the sensing mechanism should interact electronically to the rest of the signal-processing units. Usually, the sensors are realized by micromachining process of Si-wafer, whereas the signal-processing unit is materialized by standard CMOS technology. Several integration techniques are adopted globally to merge the MEMS and CMOS technologies together [10, 11].

1.1.1 Generic Process Flow of MEMS Technology

The backbone of MEMS process is the standard microfabrication technology. Several steps are common in these two technologies. Figure 1.1 pictorially explains the generic process flow of MEMS technology.

It involves cleaning of the wafer to alleviate different types of contaminants, followed by grown or deposition process as per the stress profile of the substrate; afterward various pattern transfer methodologies are adopted to realize the actual microstructures. Unique 3D-profiles are often materialized by micromachining processes. After the successful fabrication method, usually the wafers are sent to the dicing or assembly section for further processing toward the realization of packaged device or Known Good Dies (KGDs). Packaging method can be a bottleneck in the maturity of MEMS product; hence, extensive unique methodologies are adopted

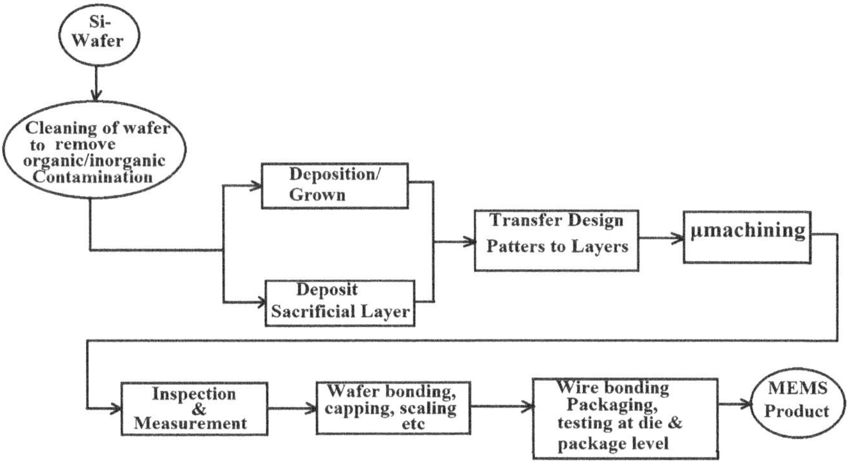

Fig. 1.1 Schematic diagram of basic process steps involved in MEMS

Table 1.1 CMOS process and subsequent layer in microdomain

CMOS process	Micro-device layer
Implantation	Piezo-resistor, electrode
Deposition/oxidation	Structural, sacrificial, masking
Polysilicon	Resistor, electrode, structural
Metallization	Electrode, membrane, conductive layer
Passivation	Structural layer, stress compensation
No such process is present here	Micromachining
Thin wafer bonding	Bonding (fusion, eutectic, anodic)

for different flavored devices. Table 1.1 briefly demonstrates some generic CMOS process steps in association with its μ-device layer. Same discussion holds true for MEMS process for realizing sensors.

Sensor parameters such as offset, hysteresis, accuracy, and repeatability are having greater impact due to contamination.

1.2 Outlines on Wafer Contamination

As discussed earlier, there may be major three types of contaminations on the silicon wafer surfaces. It may come from external particles from surrounding ambience, or generation of particles from in situ processes like residues of polymers metal, native oxide, or thin films, after pattern transfer procedures. Sometimes, it can be generated from the associated chemicals, gases, process tool, handling of wafers, parasitic reaction between materials or dissolution of tool parts, ignorance of clean-room protocols, etc. Figure 1.2 depicts the various sources of contaminations. The final reliability aspect of the μ-device is solely dependent on his contamination issue. After the crystal growth of silicon wafer during its transportation, the contamination may incur, which is tried to be eradicated by initial chemical cleaning (RCA/SC-1/SC-2) of base wafers. In situ contaminants produced from process steps, like lithography, wet/day etching, ashing, etc. are alleviated from extensive rinse/day method by applying DI(de-ionized) water, through centrifugal method. Shelf life of the chemical also plays an important role to generate unwanted impurities. Hence, strict mandate is followed in IC industry for not using any expired chemicals, which is never recommended for better yield of the whole process chain. Table 1.2 summarizes some important chemicals used in basic μ-machining process.

Impurities not only comes during process steps, but also it may occur during post-fabrication methods, like wafer dicing, assembly of bare dies in the PCB or COB, transferring of dies from one place to another places with different classes of clean room, etc.

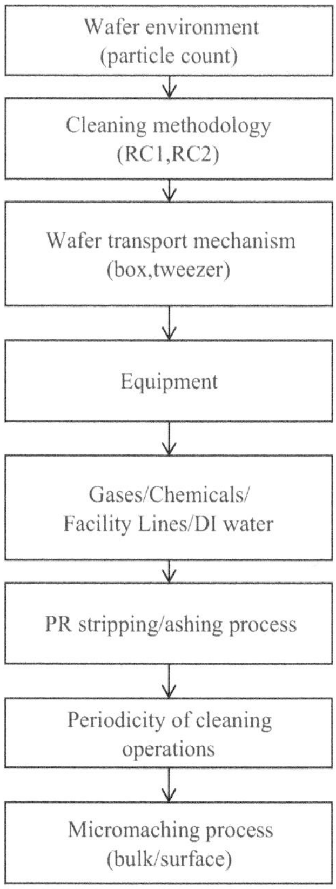

Fig. 1.2 Flowchart showing various sources of contamination

Table 1.2 Overview of generic materials employed in micro-device process

Processes	Bulk micromachining	Surface micromachining
Diaphragm/Base layer deposition (furnace)	SiO/SiN	SiO and SiN
Active/conductive layer(deposition and patterning)	Poly Si, Ti, Al	Poly Si, Al
Passivation/sacrificial layer(deposition and patterning)	SiO and /SiN	PECVD SiO, Polymide or PSG/BPSG
Metal and structural layer(deposition and patterning)	Al/Au/Poly	Al, Au,SiN, Poly
Chemistry	Wet (KoH, TMAH) DRIE based	HF, HF+Glycol,CPD, Vapour HF

In the Fig. 1.3, such two devices (MEMS/μ-machined) have been shown, with contaminations. Figure 1.3a is a bulk micromachined device part, backside of an antenna array, especially after cavity etching through KOH bath. Figure 1.3b depicts one surface μ-machined device, which is realized after removal of sacrificial layer during the process. Air-gap is clearly visible in this device part. Detailed discussion upon these devices is not matter of subject in this chapter. However, the impurities/contaminants are clearly/visible in these processed devices. Such contaminants can cause severe failure concern in the final product.

Traditionally adopted rinse and dry method is not implemented in these delicate devices manufacturing for removing such contaminants. Rather, special processes like critical point dryer (CPD) process or vapor hydrofluoric acid (VHF) etching or vapor phase cleaning procedures are utilized in such delicate or fragile 3D-microstructures. In general, commonly addressed failure mechanisms of MEMS devices are stiction, sagging due to imbalanced stress profile in the device layers, mechanical fracture, crack, wear and tear problem, humidity, and radiation inspired

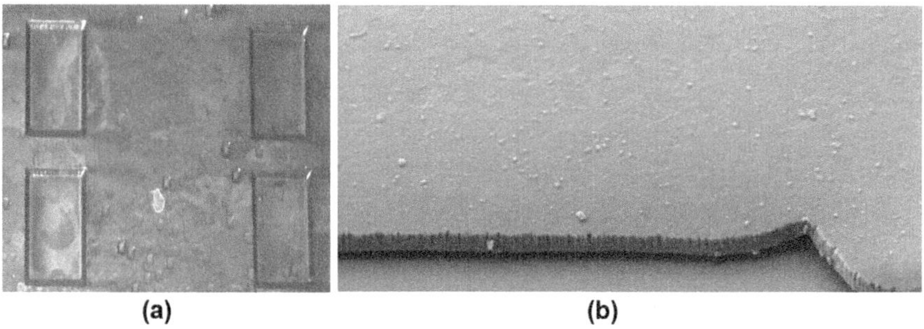

Fig. 1.3 Impurities on the processed devices. **a** Bulk. **b** Surface micromachined membran e [13]

phenomena. However, contamination-related failure mechanisms are least explored areas in such devices. Electrical shunt or open circuit may happen due to this unexpected contamination. For example, a conductive contaminant between the air-gap of RF-shunt switch structure may invite expected shunting phenomena, thus leading to change its capacitance ratio (ON/OFF), i.e., figure-of-merit of the whole device. Or, in high-frequency devices, the roughness of the metallic surface may incur extra losses, yielding in skin effect and alteration of desired characteristics. Even, an impurity between the fingers of interdigital-capacitor (IDC) structure may lead to change the whole electric-field pattern [15–17].

1.3 Methods of Cleaning Impurities

From the aforesaid discussions, it is well understood that cleaning is the basic need for any microfabrication methodology. For this specialized field of VLSI/ULSI/MEMS technology, a small particle of size ~ 0.5 μm can give detrimental effect in final operation of the device. Standard cleaning agents are employed in this μ-fabrication process, which is summarized in Table 1.3.

Properties of the substrate's upper surface may be altered with the chemical treatment during the cleaning process. A hydrophobic surface may be changed to hydrophilic and vice-versa with the chemical interaction at the interface layer(s).

Table 1.3 Cleaning agent employed in micro-device fabrication

Cleaning agent	Composition	Remarks
SC-1	$NH_4OH + H_2O_2$	Organic
SC-2	$HCl + H_2O_2$	Organic and inorganic
FSI	$NH_4OH + H_2O_2 + HCl$	Organic and inorganic
Piranha solution	$H_2SO_4 + H_2O_2$ (110 °C)	Photoresist
HF	Buffered(NH_4F + HF), Diluted& Standard HF	Dielectric/Native oxide
EKC	EKC + IPA (@65 °C)	Polymer
Ultrasonic	De-ionized(DI) water bath	Removing surface impurities
Megasonic	DI bath (megasonic frequency with variable power)	Removing surface impurities
Rinsing and drying	DI water resistance = 13–18 MΩ	To remove particulates observed in process
Plasma operation	Argon gas	Etching and ashing

1.3.1 Mitigation Techniques to Prevent Contaminations

Nucleation sites can be generated due to in-line defects or impurities, which can further diffuse into various device layers in subsequent process steps. Signature or the failure modes of the final device are reflected as a 'defect.' However, proper and accurate measurement techniques are essential to detect such ill effects. It requires extensive analytical skill of the professionals involved in this particular domain to investigate the right cause of failure. Contaminants/impurities can cause unpredictable behavior of the device. Hence, in industry, various dummy process runs are carried out before implementing actual process steps, as shown in Fig. 1.4. With these trial runs, most of the time, actual root cause of device failures is framed with probabilities.

In standard μ-fabrication process, very often the surface of the wafer is cleaned with RF-plasma at 13.56 MHz (ISM-band). Air flowing pattern, its rate, humidity and temperature of clean room, purity of gases, chemicals, vessels used for precursors, etc. all these metrics play pivotal role in the controlling of impurity level of the clean room. Even, the furniture used in such environment should be of some recommended semiconductor/clean-room grade/certified, so that it should not impact on actual fab-line process. Frequently, the dust particles are monitored inside the clean room and logged to observe the protocols holistically. Ultra-low-particulate filters (ULPA) are employed in the clean-room ambience. Proper care is also taken to ensure the vibration-free platforms. Here in Table 1.4, various standard techniques have been described to capture particulars in clean-room ambience.

Residues of impurity can result in the failure of device because of pin-hole, void, crack, and dislocation of deposited films. Traditionally, two techniques are adopted to detect the contaminations, viz. in-line and off-line. With these methods, engineers can unveil the root cause for metallic contamination, ionic impurities, chemical contaminations, etc. Several characterization techniques are available now-a-days to capture these contamination issues, like EDX, gas chromatography, ion chromatography, spectroscopy, etc. [9]

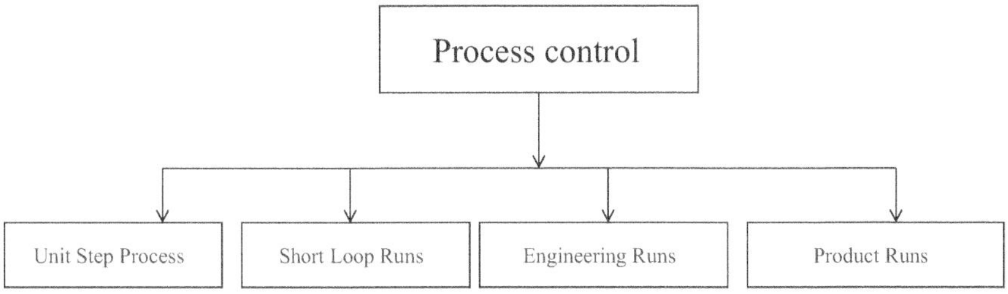

Fig. 1.4 Process control steps

Table 1.4 Particle size and measurement technique

Particle nomenclature	Sizes (nm)	Technique
Macro	> 50,000	Naked eye
Micro	100–50,000	Optical microscopy
Submicro	10–100	Optical microscopy
Nano	1–100	Electron and probe microscopes
Atomic	0.01–1	Holography
Subatomic	< 0.01	Spectroscopy

1.4 Gate Oxide Integrity

Integration of micro-devices with CMOS/interface electronics is the ultimate goal for the modern world smart sensors, and gate oxide integrity is the important aspect in this context. Contamination in the grown oxide can affect significantly the performance of the whole CMOS transistor. Thus, it is treated as prime parameter of the whole fab-line process. Presence of mobile ions in the oxide can shift the flat-band voltage of the CMOS; thus, it becomes as one of the process indicator in VLSI process. The quality of the gate oxide depends on several factors. In this section, we discuss about various controlling parameters, such as thickness, cleanliness, and process ambience, and a thorough study is conducted thereafter [12–14]. Diffusion furnaces are employed for various high temperature processes such as annealing, alloying, poly, oxide, and nitride deposition. Grown oxide quality is critical for the MOS device and circuit realization as contamination results in the threshold shift which is detrimental for the repeatability and yield. The standard process establishment consists of oxide deposition and its characterization. (Capacitance vs. voltage) measurement carried out for each trial apart from uniformity and thickness measurement on the wafer. Standard 150 mm wafer is taken for the experimentation and grown oxide of various thicknesses are analyzed. As gate oxide affects the transistor characteristics so quality of gate oxide and its contamination control are important parameters in the CMOS technology.

Oxide quality at the back-end is important for the MOSFET reliability as flat-band voltage changes due to the mobile ions. Furnaces are mostly employed to achieve high-quality oxide layer. The thickness of the oxide shrinks with the technology node (with each node 0.7 $\times$ reduction) as per Table 1.5 [9, 10].

Table 1.5 Gate oxide thickness in various technology nodes

Technology node (µm)	Typical gate oxide thickness (A°)
0.18	30
0.80	180
1.2	250
3	500

Table 1.6 Wet cleaning methodology

SC-1 cleaning	SC-2 cleaning
Particle removal	Gold and copper residues
$NH_4OH:H_2O_2$:DI:: 1:1:5 (@80 °C)	$HCl:H_2O_2$:DI:: 1:1:6 (@70°C for 5–10 min)

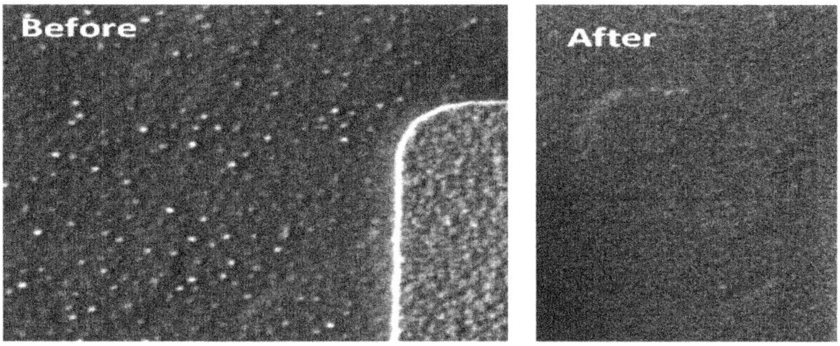

Fig. 1.5 SEM image of capacitor electrode before and after cleaning [courtesy:SCL]

As the technology node shrinks, it is imperative to have higher quality gate oxide due to smaller oxide thickness (Table 1.6). The process of oxidation using the furnaces yields high-quality oxide but controlling the number of mobile ions and resultant shift in voltage are the critical parameters which need to be evaluated for establishing the quality of the gate.

Role of cleaning and establishment of process is shown in Fig. 1.5 where polymer residues have gone after introduction of modified process.

1.4.1 Overview of Oxidation Process

Generally, oxide thickness of 500 A°/1000 A° is kept for the oxide quality measurement. Basically, chemical vapor deposition methods such as LPCVD, APCVD, and PECVD are employed for the oxide deposition. CVD films are conformal whereas plasma vapor deposition (PVD) such as sputtering or evaporation require line of sight between the surface to be coated and the source. CVD leads to high purity without the constraint of the requirement of high vacuum. The oxidation process can be segregated into the following steps-

- Transportation from the bulk of the oxidizing gas to the outer surface of silicon where it is adsorbed
- Reaction at the interface with Si to form SiO_2

The adsorption of oxidant can be written as-

$$F = h(C* - C_0), \tag{1.1}$$

where h is the gas-phase transport coefficient, C^* is the equilibrium concentration of the oxidants in the surrounding gas atmosphere, C_0 is the concentration at the oxide surface.

Variations such as gas flow rates, spacing between wafers and wafer orientation have minimal or no effect on the oxidation rates [2]. The flux representing the diffusion of the oxidants from oxide layer to the Si-SiO_2 interface can be expressed as

$$F = D(C/x) \tag{1.2}$$

The various processes of CVD are: atmospheric pressure CVD (APCVD), low pressure CVD (LPCVD), plasma enhanced CVD(PECVD), and the main process parameters in the CVD are: temperature, pressure, velocity (gas flow), and distance. The process related with the oxidation can be represented as

$$\begin{aligned} 3SiH_4 + O_2 &\xrightarrow{500\,^\circ C} SiO_2 + 2H_2 \\ SiCl_2 + 2\,H_2O &\xrightarrow{900\,^\circ C} SiO_2 + 2\,HCl. \end{aligned} \tag{1.3}$$

Table 1.7 shows the CVD processes and the features associated with each process.

Temperature is one of the key important parameters in deposition process, where high temperature can lead to desorption while low temperature leads to impurities and structural defects. In the standard process, heating elements of the tubes are baked to remove any contamination and heaters are kept at 200 °C for 12 h continuously. This procedure is performed without the tubes as it helps to develop the heater with an oxide layer which acts as a protective layer against contamination of the tubes by the heater. After the insertion of the quartz tubes into the furnace, heaters tubes are given a bake out cycle. This bake out continued for 24 h with constant nitrogen flowing through the tube. This serves for the migration of the organic impurities, toward the

Table 1.7 Furnace processes

Process	Temperature (°C)	Pressure	Layers	Feature
APCVD	~ 700–800	1 atm	Oxide/nitride	Higher throughput, poor step coverage
LPCVD	~ 600–620	0.25–2 Torr	Oxide/nitride	Higher throughput, poor step coverage, less particles
PECVD	~ 400	100–200 mTorr	Oxide/nitride	Low throughput, higher etch-rate, more particles, poor step coverage
MOCVD	~ 600–1100	30–200 mTorr	Composite semiconductor with metal organic precursors	Excellent uniformity, good conformity

Table 1.8 Experimentation and various analyses

Conditions	Oxide thickness (A°)	Voltage shift (mV)	Capacitance (pF)	C_{fb}/C_{min}	Impurities	Remarks
I	1300	600–1000	1700	0.9	10^{11}	No cleaning
II	1300	800–2000	224	0.25	4×10^{11}	Smaller area compared to I
III	1000	400–1000	235	0.278	2×10^{11}	Ultrasonic cleaning
IV	1016	90–160	250	0.282	3×10^{11}	Cleaning of furnace and C-V screen

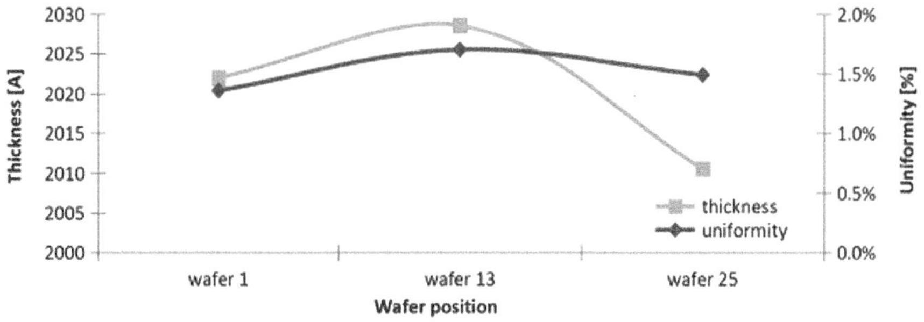

Fig. 1.6 Uniformity and thickness plot

deep inner side of the impurities resulting in protecting the future process runs from getting contaminated. It also helps to oxidize the organic impurities.

N-type single side polished wafer (150 mm) is taken having standard resistivity of < 10 Ω-cm. Wafer boat consisting of multiple wafer (wafer no # 1, 13 and 25) is taken for the process trials, and various parameters such as thickness and uniformity are measured as shown in Fig. 1.6.

The targeted thickness is 2000 A° with the average growth rate of 72 A°/min and average time taken for deposition is 28 min having cross-load uniformity of ~ ± 0.5%.

1.4.2 Experimental Set-Up

StandardN-type silicon wafer (675 ± 30 um) is taken with the resistivity of < 10 Ω-cm for the characterization. After standard cleaning, oxide deposition of 1000 A° is carried out followed by the metallization and alloying. The back side oxide is removed before C-V characterization of the wafer. The wafer is biased from +

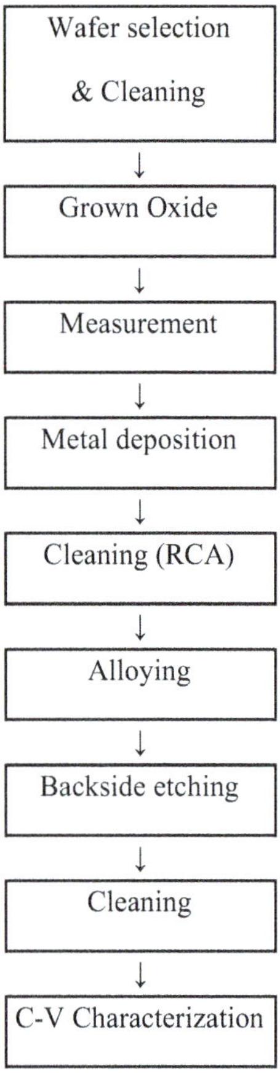

Fig. 1.7 Flow chart of characterization

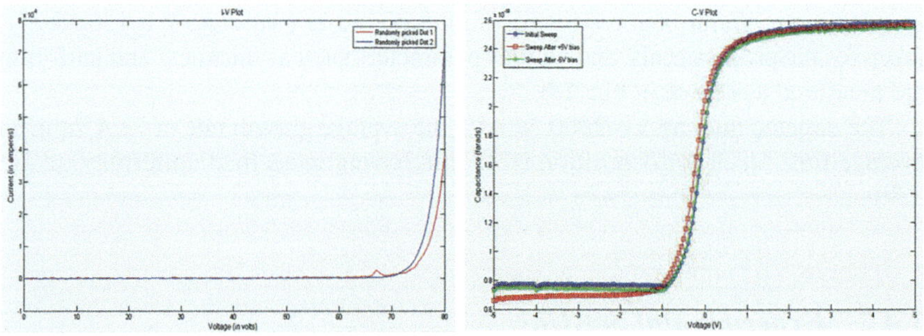

Fig. 1.8 C-V and I-V plot of the deposited oxide

5 V to -5 V and thermal stress is provided on the wafer. Various parameters such as capacitance, capacitance, voltage shift, and impurities are measured using the C-V plotter.

Ultrasonic cleaning apart from wet cleaning (NH_4OH + DI) is performed so as to further remove particles. Bias voltage along with AC signal is applied for the capacitance measurement. Multiple thermal stresses are given along with the voltage sweep and voltage shift is monitored and recorded.

The effective series resistance in all cases is < 1 mΩ. Shape of the curve, voltage shift, number of impurities, and capacitance are the important parameters to characterize quality of deposited oxide. The number of mobile ions (impurities) targeted to be in order of 10^{10} and voltage shift to be within 200 mV. The various conditions segregated as I to IV show that thickness variation, wafer type are having minimal effect on the oxide quality and other parameters such as cleaning of the wafer and equipment along with the process stability plays important role.

Main parameters affecting the gate oxide contaminations are: stability of process parameters, contaminations (equipment and wafer pod), and cleanliness associated with the characterization tool, process delay, etc.

The contamination measurement carried out with various variants using C-V plotter is tabulated as below:

Conditions IV results in the minimal voltage shift and also less impurities which is attributed to the various cleaning methodologies primarily in various tools.

Role of contaminants is important at lower technology node and process establishment of the same is important so as to control impurities and voltage shift. The summarizations of the findings are:

- Type of wafer is immaterial (p- or n-type) in the C-V measurement.
- Cleanliness of the equipment, environment, and characterization tool is important considerations.
- Measurement to be carried out between within wafer to wafer front.
- Uniformity and thickness control are important process before characterization.

Cleanliness of the furnace and the various other tools are important apart from the various cleaning methodologies.

1.5 Summarization

In modern-era communication, microfabrication technology is an emerging field in the areas of fluidics, chemical, physical, navigating, inertial, optical, wireless, and biological domains. As compared to lower-node of VLSI technology, MEMS technology having comparatively larger dimension geometrics is not that much prone to contaminators on impurities. However, the mechanical behavior of this kind of sensor may be altered significantly because of impurity imperfections.

Voids can be formed for the miniature dust particles, which may affect detrimentally the yield of the whole product. Similarly, we have also noticed that how the surface micromachined devices are sensitive to these unwanted contaminants. This chapter highlights the sources of contaminants in fab-line process, its ill effects, and finally various mitigation techniques to overcome this unwanted issue. Analytical approach, correlated with the practical data to be studied in depth for finding out the root cause.

Questions

1. What are the basic process steps involved in MEMS?
2. What are prime sources of contamination in microfabrication technology?
3. What are the different kinds cleaning agents employed in μ-device fabrication?
4. How the gate oxide thickness varies with the CMOS technology nodes?
5. What are the different types of oxidation procedures involved microfabrication technology?
6. What is 'Piranha' solution? Where is it used?
7. What are SC-1 and SC-2 cleaning methods?
8. How to remove organic and inorganic contaminates from the wafer surface?
9. How the thickness of oxide or nitride films is measured in microfabrication process?
10. What is mega sonic cleaning? When it is used?

References

1. Singh K, Nirmal AV, Sharma SV (2017) Characterization and mitigation of electro-static bonding failures in micro sensors. ICTACT J Microelectron 3(1)
2. Singh K (2017) Overview of MEMS sensors and associated aspects. ICTACT J Microelectron 3(2)
3. Singh K (2017) Contaminations in MEMS processes and removal methodology. ICTACT J Microelectron 3(1)
4. Kohli R, Mittal KL (2015) Developments in surface contamination and cleaning. Elsevier
5. Feeder GK (2003) MEMS fabrication. IEEE-ITC International Test Conference, pp 691–698
6. Gad-el-Hak M (2002) The MEMS handbook. CRC Press
7. Anathsuresh V, Gopalakrishnan B, Aatre M, (2010) Smart systems. Wiley India
8. Bryzek J (2011) Principles of MEMS: handbook of measuring system design. John Wiley & Sons
9. Kern W (1990) The evolution of silicon wafer cleaning technology. J Electrochem Soc 37(6):1887–1892
10. Sarkar CK (2016) MEMS and nanotechnology for gas sensors. CRC Press
11. Singh K, Nirmal AV (2016) Reliability aspects in RF-MEMS circuits for space applications. JETR 4(6):1–11
12. Grym J (2010) Semiconductor technologies. InTech Publisher
13. Baltes H, Brand O, Fedder GK, Hierold C, Korvink J, Tabata O (2005) Advanced micro and nanosystems, vol. 2. CMOS – MEMS - 2005 WILEY-VCH Verlag GmbH & Co. KGaA, Weinheim
14. Dealand BE, Grove AS (1965) General relationship for the thermal oxidation of silicon. J Appl Phys 12(36):3770–3778

15. Fisher G, Seacrist MR, Standley RW (2012) Single crystal growth and wafer technologies. Proc IEEE 100:1454–1474
16. Karmakar A, Singh K (2019) Si-RF technology, ISBN: 978-981-13-8051-8. Springer, Singapore, 1st edition
17. Karmakar A, Biswas B, Chauhan A (2021) Investigation of various commonly associated imperfections in Radiofrequency Micro-Electro- Mechanical System devices and its empirical modelling. IETE J Res 1–10. UK Landon. Link: https://doi.org/10.1080/03772063.2021.1880342

Chapter 2
Wafer Criteria

This chapter outlines on the various types of silicon wafers employed in the sensor fabrication, which are quite different them that of IC industry. Stringent specifications of silicon substrate are varied from application to application which is also covered in this chapter. MEMS technology very often uses double-site-polished (DSP) wafers, silicon-on-insulator (SOI) glass wafer, epi-wafers, high-resistive ($\rho > 8k\Omega$-un) silicon wafers for realizing different sensors, which are not traditionally used in standard microfabrication process.

2.1 Introduction

Gigantic growth is observed in the field of IC technology in terms of enabling multi-function capabilities along with miniaturization promise. The choice of substrate plays an extreme role in this holistic job. From the last few decades, silicon has proved itself as a better option for all VLSI engineers for fabricating multifaceted sensors, gates, memories, and obviously a complete solution for the IC industry. Other substrates are even not capable to penetrate in this field with the promising features that is offered by silicon substrate. Even the oxide and nitride formation of silicon are much more chemically stable than its other counterparts. Structure of the silicon can be: single crystal, polycrystalline, and amorphous types. Out of this single crystalline and polycrystalline silicon are widely accepted forms of silicon in standard IC fabrication technology. With an impurity level of around 1 ppb (parts per billion), generally the high purity and single crystal silicon are extracted [1–5]. The size of the silicon wafer changes with the progress in IC technology. For example, it started its journey from 2″ – diameter, then 4″, 6″, 8″, and now 12″. Quality of the silicon wafer is determined by the location from where it is cut from the ingot. Its price is also one prime factor, which depends upon this.

K. Singh and A. Karmakar, *Micro Electro Mechanical System (MEMS) Technology for Multifaceted Applications*, https://doi.org/10.1007/978-981-96-6162-6_2

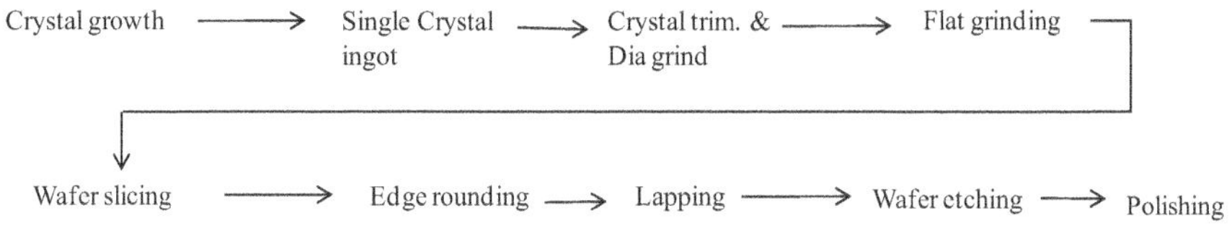

Fig. 2.1 Generic processing steps of silicon wafer preparation from its ingot

Silicon ingots are sliced and further lapped to achieve the desired thickness. Typical processing steps of wafer processing are summarized in Fig. 2.1.

The specification of the wafer is fixed upon judiciously depending upon the domain applications viz. optoelectronics, MEMS, CMOS, sensors, etc. For example, MEMS sensors demand SOI, DSP, and high-resistive silicon wafers, whereas the CMOS-ASIC application areas necessitate highly conductive or low-resistive silicon wafers. Generally, RF or high-frequency applications require high-resistivity profiles of the substrate, especially float-zone (FZ) processed wafer. Even, the RF-CMOS devices use ~ 10 Ω-cm silicon wafer for RF end usages. Epitaxial wafers are mostly used in optoelectronics devices on photovoltaic applications. Traditionally, silicon-on-insulator (SOI) wafers are well accepted for preventing the latch-up problem in CMOS and it is extensively used in realizing bulk or surface micromachined sensor's diaphragm with very tight tolerances. Table 2.1 summarizes various commonly used materials for MEMS technology [6–8].

Table 2.1 Commonly employed materials in micromachining technology

Materials	Thermal expansion coefficient 10^6/K	Thermal conductivity at 300 K (W/m-K)	Specific heat (J/kg-K)	Density (kg-m^3)	Electrical resistivity (Ω-m)
Si	2.6–3.1	150	700	2330	2330
SiO_2	0.5	1.4	1000	2200	10^{12}–10^{14}
Si_3N_4	3.3	30	1100	3270	10^6–10^{15}
SiC	3.9	111	667	3210	31.66×10^{-8}
Nickel	13	90.7	0.44×10^3	8900	6.8×10^{-8}
Pt	8.8	70	130	21,090	10.6×10^{-8}
Au	14	295	130.2	19,300	2.2×10^{-8}
Air	3430	0.024	1.0×10^3	1.239	1.0×10^{18}
Poly-silicon	2.8	29–34	702	2330	3.22×10^{-7}
Porous silicon	2	1	850	466	–

2.2 Wafer Growth Technique and Its Specifications

Single crystal growth technique of silicon wafer can be broadly categorized as CZ (Czochralski) on float-zone (FZ) process. Preparation methodologies in these two aforesaid processes are quite different in terms of its growth rate, dislocation density, distribution of dopant, etc. Mostly, CZ-process is preferred to manufacture CMOS-grade wafers, which offers excellent control over the growth rate. The FZ-technique offers high purity for realizing high-resistive silicon (HRS) wafers, ranging the resistivity from 1 kΩ-cm to 10 kΩ-cm; higher various literatures describe well the wafer growth technique and the detailed specifications of standard silicon wafer [9]. However, in this chapter, we take this opportunity to discuss more about wafers and its related information in context with MEMS technology.

Keeping in view of monolithic integration of MEMS sensors and its associated CMOS interface or application specific integrated circuit (ASIC) are targeted or adopted in now-a-days world.

Silicon is the best choice for this holistic, as compared to its other counterparts, such as GaAs, alumina. It provides a common platform for integration of active and passive modules of a complete sensors/transducer.

Important parameters of Si-wafer are: size/diameter, smoothness, bow, TTV, warpage, resistivity, etc. A tight tolerance of all the parameters dictates the quality of the wafers, and hence the economical sides of the wafer. Internally, the grain boundary decides the resistivity profile of the substrate. The type of dopants (n of p) is introduced for the specific applications. Table 2.2 briefly sketches the comparison between crystalline silicon and polysilicon in terms of few important parameters.

Either the wafer flat (major or minor or both) or a groove is employed in each wafer. Figure 2.2 shows two different types of wafers with flat and groove. Apart from the major flat, sometimes wafer possesses minor flat with different orientations, which decide the type and Miller indices [9].

Table 2.2 Comparative list of material properties

Material property	Crystalline silicon	Polysilicon
Specific heat (cal/g°K)	0.169	0.169
Thermal expansion (10^{-6}/°K)	2.33	2–2.8
Mobility (cm^2/V/s)	1500	30
Poisson ratio	0.262	0.23
Resistivity (Ω-cm)	Doping dependent	Function of grain structure

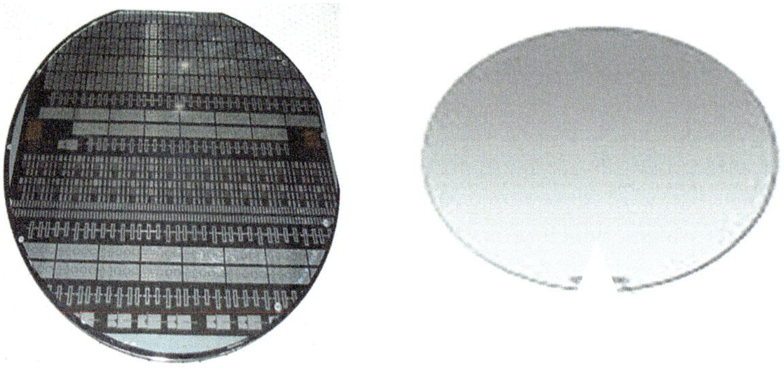

Fig. 2.2 Typical 6″ processed wafer (with flats) and 8″ raw wafer (with groove)

2.3 Different Types of Wafers for MEMS Technology

Apart from the standard low resistivity CMOS-grade wafers, various types of wafers are employed in realization of MEMS sensors. These are: HRS, SOI, and epi. Details about all these types are given in the following subsections.

2.3.1 High-Resistive Silicon (HRS/HRSi)

The main challenge of making high-frequency circuits on silicon is the substrate noise or cross-talking behavior. The substrate should be modeled properly to understand the underlying physics behind it. Analytically and experimentally both, it has been observed that the substrate loss in silicon depends strongly on two factors: frequency and resistivity. With the enhancement of operating frequency, the coupling noise due to substrate increases significantly. Effect of dielectric and surface-wave propagation becomes prominent at higher ranges of frequency. Parasitic capacitances try to affect the actual device performance. All these effect can be termed as 'dielectric' on 'substrate' loss collectively.

During the propagation of signal from one point to another point via transmission line, (here it is 'planar' type) above the substrate, the following loss mechanisms are encountered:

1. Conductor loss (due to finite conductivity of metal lines).
2. Die-electric or substrate loss (due to ideal nonzero value of loss tangent or trap charges).
3. Interface loss (due to trapped charges).
4. Radiation loss (due to uncontrolled propagation of signal at the junctions or boundaries).
5. Substrate noise.
6. Electrostatic discharge (ESD)/dielectric charging phenomena.

As mentioned earlier, the resistivity profiles of HRS are of the order kΩ-cm, which is around 100 times higher than that of standard wafers. Two other critical parameters of HRS are uniformity of resistivity throughout the thickness of wafer and radial uniformity, even during the process conditions.

2.3.2 SOI Wafer

SOI wafer can be manufactured by three ways—silicon implanted by oxygen (SIMOX), bonding wafers and zone-melt-recrystalline (ZMR) polysilicon.

It is widely used for optoelectronic devices, RF/μwave devices, pressure sensors, power, and high temperature applications. In the SIMOX procedure, after implantation of O_2, annealing is done at around 1300 °C. With the energy and dose of the implant procedure, the whole process is controlled to define the thickness of the wafer. And off course, temperature plays an important role in the whole process during annealing.

In the wafer bonding process, in one wafer, an oxide layer is grown (~1 μm) over silicon, which is further bonded with another wafer at a temperature of ~ 1100 °C in an inert gas (N_2) ambience. Mechanical grinding is finally carried out with polishing and chemical mechanical polishing (CMP) procedures.

In the ZMR techniques, polysilicon layer is deposited on an oxidized Si-wafer and recrystalline using LASER or evaporation method.

Other variants of these kinds of wafers are: Silicon-on-glass (SOG) and silicon-on-sapphire (SOS). In these kinds of wafers, traditional silicon wafer is thinned, and afterward, it is mounted to a supportive insulating layer.

Apart from the generic specifications of the wafers, choosing SOI wafer depends upon the thickness of its device and isolation layers. Figure 2.3 shows a cross-sectional view of a generic SOI wafer.

Main attractive feature of SOI wafer is the reduction of noise parameters, enhancing the noise immunity, controlling precisely the thickness profile of diaphragm in pressure sensor or any bulk micromachined devices.

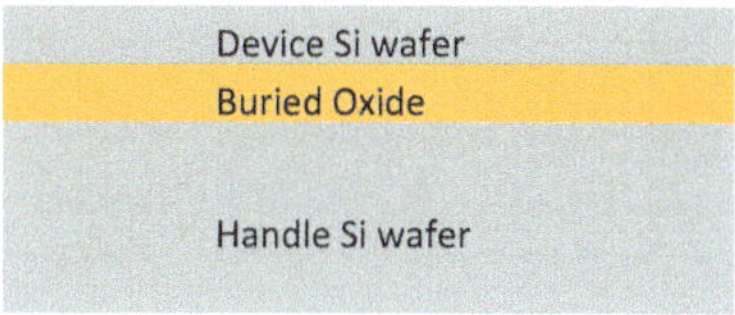

Fig. 2.3 Cross-sectional view of SOI wafer

2.3.3 *Epi-Wafer*

It is a special type of silicon wafer, where a thin, controlled thickness of layer is laid over the main substrate. In additionally, this kind of layer is grown in two ways—homo-epitaxy or hetero-epitaxy with the help of chemical vapor deposition (CVD) or molecular beam epitaxy (MBE) methods. In CMOS devices, such wafers are popularly used to improve the latch-up susceptibility parameter. The thickness of the epi-layer is a function of process time and in situ temperature. The layer can be buried or multilayer and are grown at temperature below melting point. Figure 2.4 shows a cross-sectional view of epi-wafer.

In the above three Sects. (2.3.1–2.2.3), various types of silicon wafers used in MEMS technology are explained. In the final usage, apart from specific parameters, the wafer specs are defined by SEMI standard. Day by day with the increasing demand of compact devices implementing lower geometry/node, the wafer sizes are getting larger. Table 2.3 shows such example with the weight of wafer as another spec.

2.4 Consideration of Materials for High-Frequency Applications

In the previous subsections, we have discussed about the importance on the type of wafers in the development of microstructure. Ease of availability, processing cost, technology adopted for realizing practical devices, become salient features in determining the type of material for any device fabrication. Additionally, losses associated with the substrate in terms of high-frequency application also play a pivotal role. Further, integration capability of the interface electronics with the sensor development also becomes a deciding factor for choosing right substrate.

This subsection outlines various materials useful for RF/microwave–MEMS technology, process, constraints, and probable solutions to resolve those issues.

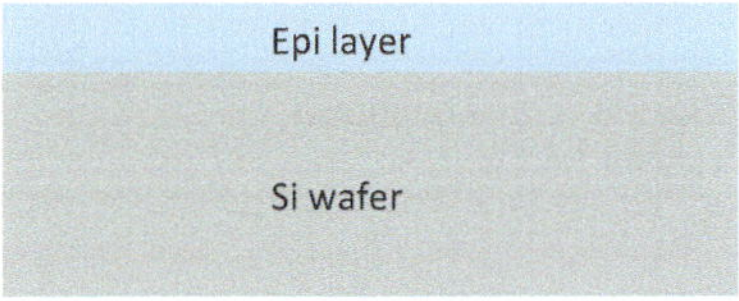

Fig. 2.4 Cross-sectional view of epi-wafer

Table 2.3 Comparative chart of wafer size vs weight

Wafer (Si) size/thickness	Weight (gm)
4″ /525 μm	9.5
6″/675 μm	25
8″/725 μm	53.4

Choice of material plays important role in the MEMS packaging, and hence, the reliability issues are greatly correlated in this regard.

Material science is the backbone of micromachining process, i.e., the development of sensors. With the rapid growth of VLSI technology, the size of the gate or transistor is diminishing day-by-day. However, this discussion does not hold true in MEMS technology, where most the sensors are based on either 0.8 μm or 1.2 μm technology. The choice of material in the MEMS depends predominantly on the application areas. For example, in Bio-MEMS application, it becomes rudimentary to use materials, which are bio-compatible in nature. Liquid crystal polymer (LCP) is such a bio-compatible, radiation resistant, substrate used for Bio-MEMS application. Substrates having low-loss tangents (tanδ), high resistivity, smoothness of the order of semiconductor grades are the generally choice of materials in MEMS processes. Usually, silicon, GaAs, sodium-rich glass, quartz, and LCP are treated as a base substrate material. Apart from using high-resistive version of the substrate, a thick buffer layer of either oxide or nitride or polymide or a stack of oxide—nitride both is generally deposited over the base substrate for realizing the MEMS devices. During this deposition, not only it is kept in mind that, we should avoid to make the metal–semiconductor function but also to manage the stress profile of the whole substrates along with the buffer layers. Silicon is always preferred as a prime choice of material in MEMS technology keeping in view of monolithic integration of active and passive devices offering a complete solution for sensor technology.

2.4.1 Selection of Substrate

Substrate properties not only mechanically decide the performance parameters of the device, but also the electrical parameters. There are several performance metrics of substrate material. We are naming few of those: dielectric constant / permittivity, loss tangent, dielectric breakdown, thermal conductivity, electrical conductivity, water absorption capacity, surface roughness, hardness, etc.

Dealing with high frequency is always an important task for professional engineers. Substrate properties play an important role in the radio frequency applications. Losses due to substrate may be incurred because of unwanted leakage current flowing through it and surface-wave-related problem may cause extra energy loss. In silicon industry, orientation of the crystal also plays an important role for further processing, such as—it may affect different etch-rate with chemicals or its affinity to some common chemical reaction. For example, in MEMS technology, generally people use < 00> , < 111> , < 110> , etc. silicon wafers. The etch-rate of < 100 > and < 111 > planes of silicon with KOH solution is completely different.

Another important parameter of substrate material is the 'radiation reinstantance' behavior. Dielectric charging phenomena may be caused due to strong radiation effects. In the radiation-induced environment, huge amount of ions on radicals may be trapped inside the grain boundaries of the substrate material. This way, an unexpected surface charge density is developed inside the material. It may invite the leakage current path inside the material. Overall performance of the device at high frequency will be disturbed this way.

Dissipation factor (DF) or loss tangent is one of the important parameter of the substrates, which plays pivotal role in RF circuitries. DF value should be ideally zero. However, its value as close to the zero value is always desired. A substrate with nonzero value of DF can be electrically thought as a leaky capacitor, i.e., capacitor with a stunt resistance. This electrical behavior of the substrate defines the theory of dielectric loss. Another part of the dielectric loss is explained by the dielectric relaxation process.

For the RF application, one of the most important parameter of the substrate is 'cut-off' frequency (f_c). It defines the upper limit of the operating frequency up to which the substrate can support the operational frequency is a strong function of resistivity of the substrate and permittivity. Mathematically, it can be defined as follows:

$$\boxed{f_c = \frac{1}{2\pi\rho\varepsilon'},} \tag{2.1}$$

where ρ and ε are the resistivity and permittivity of the silicon substrate.

2.4.2 Wafer Selection

Choice of substrate in MEMS technology is governed by its application areas, ease of processing to create 3D-profiles utilizing unique processes such as DRIE, wet etching, surface micromachining, electroplating, wafer bonding.

In the previous subsections, we have discussed in detail about the various criteria for choosing the right substrate, multiple bottlenecks in that process, etc. Important parameters of the substrate and its impact on the circuit's or device's high frequency performance have been summarized. Strong factors influencing the frequency criteria of the substrate are also outlined. Process-related constraints have also been highlighted like wet etching compatible material, etch-rate dependency on the orientation of the crystal, and so on. There is always a trade-off among all these parameter optimization. For MEMS technology, traditionally silicon, GaAs, glass, quartz, and LCP are the most common substrate names.

Application engineers judiciously choose a substrate for developing a particular sensor keeping in view of following parameters:

- Flatness of wafer.
- Resistivity of substrate.
- Loss tangent of substrate.
- Thickness of wafer.
- Diameter or size of wafer.
- Orientation of crystal.

- Dopant concentration.
- Wafer growing method.
- Warpage, bow, TTV, dislocation, etc.

Conventionally, in microwave on RF applications, usually ceramics, alumina, glass substrate, FR-4, Rogers, etc. are used widely. However, in MEMS, we are discussing about the usage of semiconductor, like Si or GaAs for RF-usages. Although these semiconductor substrates are not that much good enough in contributing lesser loss tangent (DF or tanδ), their importance is valued in view of applying unique micromachining process to achieve either cavity or free-handy 3D structures. Even, the deposition of thick buffer layer over base semiconductor substrate makes the whole base material less lossy in high-frequency uses. Table 2.4 summarizes few properties of commonly used substrate material in RF/microwave frequency range.

Selection of wafer is a crucial initiating factor in MEMS. Glass wafers enriched with sodium ions are one the prime substrate used in many sensors, like pressure sensors, RF-antenna, microheater, etc. Table 2.5 summarizes various important characteristics of glass substrates. Limitations associated with glass substrate are low mechanical strength, brittle, and poor thermal conductivity. In few RF applications, usage of glass substrate is proved to be very effective in terms of its nearly zero dissipation factor. Although, etching of glass substrate needs a precise process with experienced hands.

Table 2.4 Common properties of various substrate materials for RF

Properties	GaAs	Si	Semi-insulating (GaAs)	Semi-insulating (GaAs)	Alumina
Dielectric constant	12.9	11.7	12.9	11.7	9.7
Dielectric loss	0.003	0.0025	0.003	0.0025	0.0001
Density (gm/cc)	5.32	2.33	5.32	2.33	3.00
Thermal conductivity (W/cm-°K)	0.48	1.45	0.48	1.45	0.37
Resistivity (ohm-cm)	–	–	10^7–10^9	10^3–10^5	10^{11}–10^{14}
Electron mobility (cm^2/V-sec)	4300	700	–	–	–
Saturated electron velocity (cm^2/sec)	1.3×10^7	9.0×10^6	–	–	–

Table 2.5 Glass substrate characteristics

Type	Permittivity (ε_r)	tanδ	TCE (ppm/°C)	Thermal conductivity(W/m/°C)
Ceramic glass	6.0	0.0050	8.6	0.75
Pyrex 7740 (Corning 7740)	4.6	0.003	3.25	1.13
Fused quartz	4.0	0.0001	0.5	1.58
Borosilicate (Corning 7050)	4.3	0.0047	3.3	1.1.4
Soda lime	6.0	0.02		0.94

2.4.3 Performance Evaluation of RF Devices

Traditional silicon wafers used in CMOS fabrication process are quite lossy for high-frequency applications. Literature survey reveals that the attenuation of signal in standard CMOS-grade silicon is around 17 dB/un, whereas this value is reduced to almost less than 1 dB/un for high-resistive silicon wafer. Even these loss criteria can further be reduced to 0.1 to 0.2 dB/un implementing buffer layer over HRS-wafer. As already discussed, process engineers used to follow the protocol of having insulation layer of oxide or nitride or a combination of both or polyamide over base substrate material to reduce the line losses by implementing two in-line physics viz. inhibiting the formation of metal–semiconductor junction and balancing the compressive–tensile stress profile of overall device. Table 2.6 outlines a comparative analysis of various standard wafers / substrates used for RF / microwave applications.

Table 2.6 Comparison of transmission line losses over various substrates

Substrate	Losses (dB/cm)	Remarks
Std silicon	17 (1–35 GHz)	Higher losses associated with bulk substrate
HRSi with insulating layer	1(1–35 GHz)	Interface and other losses associated with oxide layer
Direct metallization on HRSi	0.1–0.2(1–35 GHz)	Formation of Schottky contact, thin film adhesion
Glass	< 2.5 (X-band)	Loss variation, thin film adhesion
Alumina	< 0.1	Hybrid approach for RF-CMOS integration

2.5 Concept of Micromachining

Two popularly known approaches are adopted in MEMS/micromachining process, as bulk and surface micromachining. Major advantage of bulk micromachining process is to enabling the removal of bulk of the lossy substrate material beneath the actual high-frequency devices, and thus to make it almost free from surface-wave and leaky-wave effect. Dielectric loss is reduced with this method significantly and people have reported various high-frequency structures realization even up to W-band (75–110 GHz) implementing this method. Realizing aperture-coupled antenna or high-Q cavity filter or implementing 3D-vertical integration of multiple bonded wafers is some of the important way outs with this process. Enhancing the radiation efficiency and gain of low-profile patch antennas are becoming feasible with the implementation of this unique method [10–13].

On the other hand, surface micromachining process is much popular for realizing cantilever or fixed–fixed beam profile of RF-switch structures. Even, this process has paved the way to realize high-Q induction structure in IC fabrication method. Here in Table 2.7, a comparative study is sketched to bring out the difference between these two popularly used processes [14–20].

Base silicon wafer can be etched isotropically or anisotropically with the help of chemical means or physio-chemical reactions. Hydrofluoric acid, nitric acid, and acetic acid (HNA) solution is used for isotropic etching of Si-wafer, whereas KOH, EDP, TMAH, hydrazine, etc. are used for anisotropic wet etching. In the DRIE process, silicon etching is done with an aspect ratio of 22:1 with $90^\circ \pm 2^\circ$ profile angle. Generally, standard 'Bosch-process' is used for this purpose.

Table 2.7 Micromachining processes (comparative study)

Bulk micromachining	Surface micromachining
Typically uses wet etching techniques or DRIE-Bosch process	It uses dry etching techniques like-VHF, CPD process, etc
Suitable for simple geometry such as thermal sensor,micro machined neuron wells	Suitable for complex geometry such as micro valves and actuators(free-hanging structures)
It causes a well known phenomenon in the micro electronics industry called undercut	It does not cause undercut problem. However, earlier process used to invite stiction problem
Sacrificial layer is not required	Etching of sacrificial layer is required
Limited to low aspect ratio in geometry as surface dimensions are much greater than depth as height is limited by thickness of silicon wafer	Not constraints by thickness of silicon wafer,so high aspect ratio geometries can be fabricated using surface micromachining
Larger dimension structures are made with bulk micromachining	It is difficult to build larger structures with this process
This creates structures inside a substrate	This creates structures on top of a substrate or layer of substrates

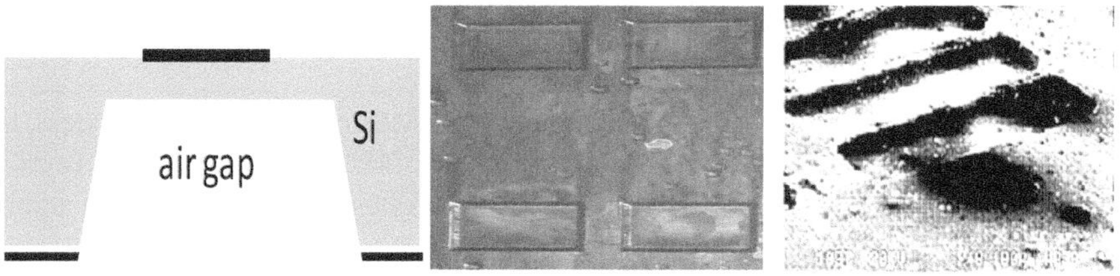

Fig. 2.5 Micromachining structure in silicon and GaAs substrate

Table 2.8 Substrate material and subsequent etchants

Substrate material	Wet chemistry	Remarks
GaAs	$H_2SO_4/H_2O_2/H_2O$	Anisotropic with roughness
Glass/quartz	HF	Isotropic with roughness
Silicon/ HRSi	KOH/TMAH	Anisotropic with smooth surface

Among all the wet etchants, KOH solution is the most widely used chemical for MEMS technology because of its very less detrimental effect as compared to its other counterparts like TMAH or EDP, as both of the chemical fumes are quite carcinogenic in nature. 40% concentrated KOH solution at 80 °C provides an etch-rate of ~ 1.1 to 1.2 μm/min for silicon. A continuous stirring process is essential to maintain the uniform concentration profile of the solution bath. It provides a truncated pyramidal profile with sidewall angle of ~ 54.7°, as shown in Fig. 2.5. Basically, it is time-based etching technique. Suitable etch-stop techniques are adopted to stop the silicon etching at desired thickness. Either with electrochemical means or with the formation of p–n junction, normally the etching is stopped during the process. TMAH offers almost half of the etch-rate (~0.6 μm/min) as compared to KOH solution [21–25].

Another important process is the etching of glass or oxide. It is a very critical process, which needs extensive care and control because hazardous hydrofluoric (HF) acid is used in this process. Table 2.8 summarizes some common substrate's etchants used in MEMS process.

2.6 Summarization

This chapter details about the various types of wafers used in MEMS technology, its specifications, material solution criteria, and finally important parameters governing the high-frequency applications. Discussion was chalked out about the various commonly employed substrate materials, like silicon, HRS, GaAs, glass, etc. The concept of micromachining along with its real-time application areas was also covered with pros and cons of many etchants. Provision of adopting monolithic integration with silicon technology for developing smart sensors is also outlined in this chapter. Speciality of introducing SOI, SOS, and epi-wafers in the micro-fabrication process is also briefly sketched. Factors determining the high-frequency performance of the circuit are also described with insight physics.

Questions

1. What are the different steps involved in the preparation of standard silicon wafer?
2. How do you differentiate between standard silicon and polysilicon material?
3. Name few wet etchants of silicon material?
4. What is HNA? What are its uses?
5. How the high-resistive silicon wafers are produced? How is it differing from standard silicon wafer?
6. Name the wet etching characteristics for the following substrates: GaAs, glass, silicon
7. What are the different types of glass substrates used in MEMS process? Make a comparison table, containing their characteristics in terms of various material properties.
8. What are the prime characteristics of a wafer used in standard microfabrication technology?
9. What are the parameters of a substrate or dielectric material, which decide its cut-off frequency of operation? How they are related?
10. What are the different types of micromachining process? List out in terms of their various characteristics?

References

1. Singh K, Nirmal AV, Sharma SV (2017) Characterization and mitigation of electro-static bonding failures in micro sensors. ICTACT J Microelectron 3(1)
2. Singh K (2017) Overview of MEMS sensors and associated aspects. ICTACT J Microelectron 3(2)
3. Singh K (2017) Contaminations in MEMS processes and removal methodology. ICTACT J Microelectron 3(1)
4. Roy JN, Zarabi MJ (1997) Materials for VLSI process. IETE J Res 43(2&3):207–213
5. Reyes AC, El-Ghazaly SM, Doml S, Dydyk M, Schroder (1994) Silicon as microwave substrate. IEEE MTT-S Digest 1759–1762
6. Yang R-Y, Huang C-Y, Su Y-K, Weng M-H, Wu H-W (2006) Loss characteristics of silicon substrate with different Resistivties. Microwave Opt Technol Lett 48(9):1773–1776
7. Singh K, Nirmal AV (2016) Reliability aspects in RF-MEMS circuits for space applications. J Eng Technol Res 4 (6):1–11
8. Singh K, Bhatnagar D, Pal S (2007) CAD analysis of microstrip lines using micromachining techniques. High Freq Electrons, 30–35
9. Wolf S (2000) Silicon processing for VLSI era : process technology. Vol.1, Lattice Press publication

10. Singh K, Bhatnagar D, Pal S (2008) Characterization of RF transmission lines on ion-implanted CMOS wafers. High Freq Electron, pp 32–36
11. Singh K, Pal S (2006) Micromachining concept on gaas and a mm-wave oscillator example. High Freq Electron, Oct 2006, pp 40–44
12. Sarafis P, E Hourdakis, Nassiopoulou AG (2013) Dielectric permittivity of Porous Si for use as substrate material in Si-integrated RF devices. IEEE Trans Electron Dev 60(4):1436–1443
13. Singh K, Nagachenchaiah K (2008) Effect of resistivity variations on the performance of lossy micro-machined resonators. Microwave J
14. Karamakar A, Singh K (2013) Design and development of x-band planar Balun on silicon. IETE J. Res 510–514
15. Reyes AC, El-Ghazaly SM, Doml S, Dydyk M, Schroder DK (1994) Silicon as a microwave substrate. IEEE MTT-S Digest, pp 1759–1762
16. Burghartz J (1997) Silicon RF technology—the two generic approaches. In: 27th European solid-state device research conference, Stuttgart, Germany, pp 143–153, https://doi.org/10.1109/ESSDERC.1997.194388
17. Tilmans H (2003) RF-MEMS: materials and technology, integration and packaging. MRS Online Proc Libr 783:66. https://doi.org/10.1557/PROC-783-B6.6
18. Sarkar CK (2016) MEMS and nanotechnology for gas sensors. CRC Press
19. Fisher S (2012) Standley Silicon crystal growth and wafer technologies. In: Proceedings of IEEE, vol 200, pp 1454–1474
20. C R M Grovenor, Microelectronic material, IOP Publishing Ltd,1989
21. Micro and Smart Systems, G K Ananthsuresh et al, Wiley India
22. Singh K (2017) Substrate material considerations in MEMS processes for RF applications. Nano and Microsystems Technol. 19(7):428–431
23. Breed G (2006) A comparison of RFIC fabrication technologies. High Freq Electron 52–53
24. Yang R-Y, Hung C-Y, Su Y-K, Weng M-H, Wu H-W (2006) Loss characteristics of silicon substrate with different resistivities. Microw Opt Technol Lett 48(9), 1773–1776
25. Benaissa K, Yang J-Y, Crenshaw D, Sridhar S et al (2003) RF CMOS on high resistivity substrates for system-on-chip applications. IEEE Trans Electron Dev 50(3):567–576

Chapter 3
Micromachined Sensors Overview

In this modern era of sensors technology, there is hardly any corner, which has not touched by the wings of MEMS technology. Starting from ground-based application to air-borne domain or consumer electronics to strategic applications, MEMS has spread its aura everywhere. In this chapter, we will discuss about various process aspects of developing MEMS sensors, its associated challenges, mitigation techniques, and finally the packaging aspects of sensors to deploy in practical applications.

3.1 Introduction

Nowadays, MEMS technology is rapidly making stride in various domains. MEMS being an offshoot of standard microfabrication technology are becoming popular for sensor's world as day-by-day demands are increasing exponentially for the Internet of things (IoT) or 5G communications. Wireless sensor network (WSN) is solely dependent on various microsensors, connected wirelessly to provide the best connectivity for the end points. In recent times, most of the sensors are based on MEMS technology. Backbone of this MEMS process is the standard CMOS fabrication methodologies. Taking the advantages of precise device dimension tolerance control and batch processing, MEMS is a promising technology for developing numerous sensors within a stipulated time frame [1–3].

The micromachining process is an interdisciplinary subject, which involves physics, electronics, chemistry, mechanical, electrical, and many more. Art and science are blended here in an engineering approach. Material science is also an important aspect in this new field of engineering. Present-day miniaturized sensors are being used in multiple home appliances, mobile phones, laptops, automobiles, game-sets, electronic toys, industrial equipment, high-end strategic machines, satellite communications, navigation units, micro or pico satellites, etc.

K. Singh and A. Karmakar, *Micro Electro Mechanical System (MEMS) Technology for Multifaceted Applications*, https://doi.org/10.1007/978-981-96-6162-6_3

Table 3.1 Comparative study between standard microfabrication (IC) and MEMS

Integrated circuits	Microstructure
Typically flat structure (2D)	Generally not flat (3D)
No moving part	Moving and non-moving both
Insensitive to the environment	Sensitive to external environment
Depends on buried layer	Surface effect devices
Highly dense with multiple functionality (IC tester)	Typically perform single function (test instruments)
On chip compensation	Generic off chip compensation
Sensitive to radiation effect	Mostly insensitive, however 'dielectric breakdown' and all communication occur as post effect of radiation

They are becoming indispensable for the compact, miniaturized form factors of the multifunctional devices [4–7].

Silicon is the most widely used material in this holy task, because of its several unique features. Electrically as well as mechanically, it is best suited for this MEMS industry [8]. Already it Chap. 2, we have discussed about the various meritorious features of silicon substrate as compared to its other counterparts. Keeping in view of all those salient features, silicon is widely accepted for micromachining technology. Although there are some differentiating characteristics of this special wing of technology, it is aligned with standard IC fabrication, as summarized in Table 3.1.

Apart from the standard silicon wafers, MEMS process demands DSP, SOI, epi, HRS-type silicon wafers, and also SOI, SOS, and quartz wafers [9]. Various special etching chemicals, gases are used to achieve unique processes of micromachining.

3.2 Process Requirement

In general, microfabrication process involves deposition / growth of either dielectric or metal thin films over the surface of wafer. Very commonly used dielectrics in this process are silicon oxide and silicon nitride. And, as a metallization layer, aluminum, gold, platinum, nickel, cobalt, and doped polysilicon are widely used either with the help of DC sputtering method or E-beam evaporation technique or electroplating method. Otherwise, electroplating technique is adopted only when very high thickness profile (~20 to 30 μm) is required for only device. During all these metallization process, always some seed layers are deposited over the base surface for better adhesion purposes. Thin film growth or deposition process depends on several process parameters, and its quality is defined by several performance metrics, such as surface roughness, uniformity, coverage profile, stress behavior, layer adhesion

Table 3.2 Typical etch-rate

Deposited layer	Typical etch-rate (Å/s)	Etchant
Oxide (LPCVD)	20	CF_4, $CF_4 + O_2$ SF_6, CCl_2F_2 BOE, H_3PO_4
Nitride (LPCVD)	30	
Oxide (PECVD)	120	
Nitride (PECVD)	90	

quality, etc. For dielectric material deposition, either TEOS or silane (SiH_4) chemistry is employed. Here also, step coverage plays an important role along with adhesion quality of the oxide or nitride thin films. Thermal stability, trap-charge density, planarization capability, etc. also play important roles in determining the quality of the whole process.

Chemical vapor deposition (CVD) is a very popularly established process for dielectric deposition. It can be broadly categorized into three types: Plasma enhanced chemical vapor deposition (PECVD), low pressure chemical vapor deposition (LPCVD), and atmospheric pressure chemical vapor deposition (APCVD).

PECVD utilizes plasma to enhance chemical reactions at low temperature for temperature sensitive substrates and a wide range of multifunctional applications. LPCVD devices implement chemical reactions through high temperature environments for high temperature resistant substrates and high performance applications. Further in the APCVD process, reaction in the chamber takes place at atmospheric pressure. APCVD process has a very poor step coverage issue. LPCVD suffers from low deposition rate and PECVD has a risk for chemical contamination. Here in Table 3.2, we summarize different dielectric thin film grown/deposited through PECVD/LPCVD processes and its etch-rate variation.

Etch-rate of the thin film depends upon two prime factors—density of the thin film and its chemical composition.

Microfabrication process is based upon few basic steps viz. growing or deposition of the thin film, photolithography, patterning, wet or dry etching, micromachining, etc. Table 3.3 briefly sketches about the tools used for these major process steps.

Table 3.3 Various tools for process characterization

Measurement tools	Principle	Mode
Film thickness	Elliposometer	In-line
Film stress	Stoney's formula (light angle)	In-line
Optical microscope	Light reflection	In-line
SEM	Electron scanning	Off-line
Surface profiler	Interferometer	In-line
Current–voltage	Modulated wave	Off-line
Resistance	Sheet resistivity	In-line
Reflectance	Infrared spectroscopy	In-line

3.3 Classification of Sensor

In general, three basic microstructures are widely used in almost all MEMS sensors.

These are: diaphragm, cantilever/fixed-beam structure, and anchor. According to the application domains, the whole MEMS field can be classified into several subdomains viz. RF-MEMS (for using MEMS in RF/microwave field), Bio-MEMS (implementing MEMS in bio-medical engineering), optical MEMS or MOEMS, physical sensor like pressure, temperature and humidity sensors, inertial navigation unit (INU) using MEMS like-accelerometer, gyroscope, inclinometer, etc. MEMS is an interdisciplinary field, which is nowadays used extensively in almost all corners of human life like consumer, automotive, satellite, strategic, defense, high-frequency applications, RADAR, bio-medical field, IoT, mobile communication, wireless applications, etc. Its usage is exponentially increasing from ground-based application to airborne devices, even in very precise, accurate navigation units. With all of promising features, this technology is becoming now an important back-bone of modern-era sensor technology. Table 3.4 shows some of the glimpse of MEMS sensors, used for modern-era multifaceted applications.

Broadly, MEMS sensors can be classified as active or passive. strain gauge, thermistor, resistance temperature detector (RTD), is examples of active sensors due to its main principles of measurement, which is based upon the excitation of external current, whereas, the thermocouple, photodiode, and piezo-electric devices are of passive kind. These sensors can generate electrical signals such as thermo-electrical voltage; photocurrent and electric signals without any external voltage/current exit ate. Table 3.5 summarizes some important sensors implementing MEMS technology.

Literature survey reveals that there is many other classification types of MEMS sensors apart from mentioned in the above paragraph. Depending upon the sensing mechanism, it can be classified as piezo-resistive, piezo-electric, capacitive, and resistive types. Every category has its own merits and demerits. Based upon the available fabrication facility, technology node, and off course the final or end application, all sensors are categorized and its right sensing mechanism is adopted. Capacitive sensors are best for electrostatic actuation method, whereas resistive sensors are very easy to process the signals. However, capacitive sensors involve unwanted parasitic capacitances, nullifying which is really difficult. Pressure sensor, impact

Table 3.4 Common microstructure applications

Accelerometer	Air bags technology
Digital micro mirror	Projection display
Inkjet printer head	Printers
Pressure sensors	MAP for automobiles
Gas sensors	Various gas detection
Bolometer	Night vision cameras
Miniaturized PCR	Oligonucleotide
Oxygen/glucose sensors	Portable blood analyzer

Table 3.5 Common sensor classification and electrical parameters

Sensors	Type	Main parameters
Physical	Temperature Humidity Pressure	Sensor range, sensitivity, TCR, nonlinearity, offset, hystersis, TCS,TCO,TCR
Inertial	Gyroscope Accelerometer	Bias instability, angular random walk (ARW), sensitivity, rate random walk (RRW), bias drift, accuracy, range, resolution, cross-axis sensitivity, natural frequency
Radio frequency	Switch Phase shifter Varactor	Center frequency, insertion loss, return loss, switching speed, tuning voltage/power handing capability, bandwidth

sensors, etc. demand highly piezo-resistive mechanism, whereas the acoustic sensor relies upon piezo-electric methodology, capacitive sensing mechanism invites nonlinearity problem, which is inherent to this sensing method. However, it is free from temperature-variation effect, whereas although resistive sensors also seem to be simpler in construction, it suffers greatly from noise incurred by temperature effect. Keeping in view of all the pros and cons, one should judiciously choose the sensor fabrication method for realizing a reliable device. Table 3.6 shows that how we should emphasize on some prime or basic specification sensor or determining end usages.

The above table does not include all the parameters, like Allan-Variance graph metrics, radiation-induced parameters. Few parameters are application specific, such as angular random walk (ARW), bias instability, rate random walk (RRW), quantization noise, etc. which can be derived from the Állan-Variance curve, very much useful for accelerometer on gyroscope sensors. It is basically a statistical tool for identifying various noise types that exist in a signal. Similarly, some radiation-induced parameters, like total ionization dosage (TID), on dielectric charging phenomena, etc. are also becomes often important evaluation metrics for the MEMS sensors.

3.4 Challenges in Microstructures

One of the major reliability concerns of microstructure is the in-line residual stress. These micro-devices are mainly formed by thin films. Residual stress may be inside these films, which can further create crack, void, poor adhesion, and even the breakage of the wafer. Several factors are there for developing such residual stress such as deposition process, nanostructure of the grown films, thermal condition of the fabrication process, RH-content of the in situ ambience, etc. This stress can be of two types: tensile or compressive. Depending upon the curvature of the film or wafer, one can estimate this residual stress with the help of popularly known Stoney equation. However, MEMS sensors undergo following challenges while its realization:

Table 3.6 Sensor parameters and description

Sensor parameters	Generic description	Measurement/ comparison	Example (accelerometer)
Transfer function	Functional relationship between physical input signal and electrical output signal	…	
Sensitivity	Minimum input of physical parameter that will create a detectable output change	Volts/Kelvin, mV/ V/mmHg	mV/g
Dynamic range	Total range of input physical signal that may be converted to electrical signal	Kelvin, Pascal, Newton	± g
Accuracy	Maximum difference between actual and ideal output signals	Kelvin (% of FSO)	
Hysteresis	Same output is not achieved with the input stimulus	Kelvin(% of FSO)	
Offset/drift	Difference between the actual output value and the specified output value under some particular set of conditions		
Nonlinearity	Deviation of linear transfer function over the specified dynamic range	% of nonlinearity	% of the FSO
Noise	Distributed across the frequency spectrum and same at all frequencies	Volts/ $\sqrt{}$Hz (noise density)	µg/$\sqrt{}$Hz
Resolution	Detectable minimum signal fluctuation	Signal/ $\sqrt{}$Hz	µg/$\sqrt{}$Hz
Bandwidth	Frequency range correspond to upper and lower cut off	Hz	………
Temperature Coeff (a) Offset (b) sensitivity	Sensitivity change with temperature	%/°C	(a) mg/°C (b) ppm/°C
Response time	Time required for a sensor output to change from its previous state to final settled value	time	Sec

1. Material property of the thin film (such as Young's modules, Poisson's ratio, etc.).
2. Thermal properties, such as thermal conductivity, TCE/CTE.
3. Mass transport or fluidic property.
4. Integration with interface electronics or CMOS devices.
5. Assembly, packaging, and testing of individual sensors in working condition.
6. Hermetic sealing.
7. Process-induced phenomena, like stiction, dielectric charging, etc.

3.4.1 Assembly and Packaging Aspects of MEMS Sensors

Sensor packaging, particularly for MEMS devices, is a critical aspect of modern electronics industry. Sometimes, the packaging aspect becomes a bottleneck to be maturity of this technology. Very large percentage of total cost of MEMS device is invested in the packaging field. The main difference between IC packaging and MEMS packaging is that MEMS packing is almost always application specific, and it is greatly affected by environmental factors, handling techniques, die attach processes, and lid sealing. Even, sometimes MEMS devices that are functional in wafer form can be rendered in openable after packaging. Die-attach materials are the key players in MEMS packaging. Hard die attach solder can create high stresses, while low-stress epoxies can be high out-gassing. Hence, judiciously the material is chosen. Most MEMS devices demand hermetically sealing. To date, most of what is known about MEMS packaging remains proprietary and published literature is scarce. Along with the design of sensor, engineer should consider also design of suitable package for its. The analysis is to evaluate the electrical, mechanical, thermos-mechanical, thermos-electric property, and even the high-frequency behavior. Suitable CAD models should be extracted and it should be shared with the sensor designer; otherwise, the final package product may show some different characteristics unless it is expected [10].

3.4.1.1 Packaging Aspect in RF-MEMS Devices

Apart from the common issues addressed in subsection 3.4.1 regarding packaging aspect of MEMS sensors, RF devices need some extra precautions. Here, the cavity inside the package can also play an important role. Its dimension supports some EM-wave propagation on the cavity should be simulated for 'Eigen-mode' consideration. Even, the wire bond used within the package can also act as an antenna (or radiative) element and the spacing between adjacent bond wires are also very important to consider the cross-talking effect. Line width, spacing should conform to standard 50Ω transmission line behavior mitigating extra spurious radiation inside the package cavity. Material properties, such as loss tangent (tan δ) value and permittivity (ε_r) also play pivotal role for RF circuits. The epoxy used should be analyzed for high-frequency capability. Provision of input and output connection is very important in such package, because at this junction the whole signal may be lost or faded significantly if proper care is not taken. Usually, wafer-level packaging or chip-scale packaging or zero-level packaging are adopted for RF-MEMS devices.

3.4.1.2 Packaging Aspect in Inertial Sensor

Packaging of MEMS-based inertial sensors (accelerometer, gyroscope, inclinometer, etc.) remains a critical challenge. These displace–sensitive devices are known to be very sensitive to package-induced stresses. Devices like 'gyroscope' need special kind of packaging called vacuum packaging for its high Q-factor. Q-factor is the figure of merit in such sensor. Quality of the achieved vacuum inside the package dictates the performance of sensor predominantly. Another important parameter of packaging such inertial sensor is the coefficient of thermal expansion (CTE) mismatch among different materials involved in the packaging such as die attaches, chip on board or printed circuit board (COB/PCB) material, silicon, etc. The CTE mismatch may incur unwanted stress effect of such delicate and sensitive inertial sensor. Even the out-gassing/degassing of PCB materials of such sensor may impact its performance if it is used in space or some remote place, where hardly any change of re-turning at of device [11–14].

3.5 Summarization

This chapter details the overview of micromechanical sensors. Its fabrication aspects, material consideration, uniqueness in comparison with standard IC fabrication method, various process requirements, classification of sensors, its intricacies and packaging aspects are covered in this chapter. Various parameters defining the sensors metrics are discussed in terms of their importance in practical domains. Challenges faced by practical professionals in this specialized domain of engineering are also briefly sketched. Ways of mitigate those issues were also highlighted for realizing a reliable product. Bottleneck of this MEMS technology is the packaging. Hence, it is also discussed in detail with zeal-life device concerns.

Questions

1. What is zero-level packaging? Where is it used?
2. What is the use of vacuum level packaging in MEMS?
3. What are the challenges faced during realization of MEMS devices?
4. What is the various process tools used in standard microfabrication technique for characterization purpose?
5. What are the typical etch-rate of various types of silicon oxides and nitride using wet and dry techniques?
6. Name few types of MEMS sensors used as physical sensor, inertial unit, RF-domains
7. What are the basic specifications of a MEMS sensor?
8. What is 'hysteresis' of a micro-device? How can it be changed?
9. What do you mean by 'dynamic range' of a sensor? Is it different from 'full-scale output'?
10. What do you mean by 'offset' of a microsensor? Explain with the same example.

References

1. Anathsuresh V, Gopalakrishnan B, Aatre (2010) Micro and smart systems. Wiley India
2. Roy S, Sarkar CK (2016) MEMS and nanotechnology for gas sensors. CRC Press
3. Singh K Substrate material considerations in MEMS processes for RF applications. Accepted for journal publication
4. Singh K, Nirmal AV (2016) Reliability aspects in RF-MEMS circuits for space applications. JETR 4(6):1–11
5. Singh K, Nirmal AV, Sharma SV (2017) Characterization and mitigation of electro-static bonding failures in micro sensors. ICTACT J. Microelectron 3(1)
6. Singh K (2017) Overview of MEMS sensors and associated aspects. ICTACT J Microelectron 3(2)
7. Singh K (2017) Contaminations in MEMS processes and removal methodology. ICTACT J Microelectron 3(1)
8. Saxena A, Behera SN, Singh K, Jain P (2013) Process optimization and role of process parameters for repeatable metal etching. In: National conference on micro and Nano fabrication, CMTI, Bangalore
9. Osiander R, Darren MAG, Champion JL (2005) MEMS & microstructures in aerospace applications. CRC Press
10. Bryzek J (2011) Principles of MEMS: handbook of measuring system design. John Wiley & Sons
11. Yazdi N, Ayazi F Najafi K (1998) Micromachined inertial sensors. In: Proceedings of IEEE, vol 86, no 8, pp1640–1657
12. Grym J (2010a) Semiconductor technologies. InTech Publisher
13. Gad-el-Hak M (2002) The MEMS handbook. CRC Press
14. Malshe P, Brown WD, Eaton WP, Miller WM (1999) Challenges in the packaging of MEMS. Int J Microcirc Electron Packag 22(3), 233–241

Chapter 4
Microstructure Fabrication

Microstructure fabrication employs standard equipment apart from machining tools such as DRIE, wafer bonder, KOH bath, VHF tool. The process steps are having similarity with the standard IC fabrication process. The various individual process steps and its integration aspects are looked very critically as per the specifications. Apart from metrological tools, wafer characterization also to be carried out and the reliability aspects to be examined for the application. Wafer encounters various process which demands multiple times processing from the same equipment for deposition or etching. Microstructure fabrication process is similar to the integrated fabrication and the main process difference relates with the micromachining process. Main process steps are metallization, deposition, lithography, and etching, bulk machining which can be either using dry or wet techniques.

4.1 Introduction

Microstructure fabrication can be segregated into various steps as below:

[A] Deposition

Thin film deposition is one of the main steps in the micro fabrication. The deposition of oxide, nitride, dielectric layers, and metal layers constitute the deposition where they are employed for various interconnections or developing active or passive circuits.

[B] Doping

Changing the resistivity of the selective surface dopants is implanted using the thermal diffusion or ion implantation methodology.

K. Singh and A. Karmakar, *Micro Electro Mechanical System (MEMS) Technology for Multifaceted Applications*, https://doi.org/10.1007/978-981-96-6162-6_4

[C] Lithography

Optical sensitive film is used to transfer pattern on to the film. The mask is made of glass plate which is coated generally with chromium.

[D] Etching

Dry or wet etching techniques are employed to remove the deposit layer or etching out the bulk material. Masking material can be oxide, nitride, or hard baked resist. The choice of material is dependent on the application.

4.2 Deposition

Thin films in the integrated circuits are mainly used in the conductor and insulator film. Metallic film is used for the low resistivity sensing and the metal–semiconductor junction. Metallic film deposition is mainly carried out using the physical vapor deposition and chemical vapor deposition. Dielectric films are mainly oxide and nitride which are used for insulator and passivation layer. Other various films-thin films such as PZT, BST and plastic layer PE, PVC, PMMA, and PDMS are also used for this purpose. Materials in the deposited layers consist of the gaseous plasma and solid, and the these layers are characterized by grain size thickness, planarity, adhesion, and resistivity measurement.

Deposition is main process in the wafer fabrication and the oxide deposition is the first step in this process and furnaces are primarily employed for this purpose. The transportation mechanism is having following steps:

(1) It is transported from the bulk of the oxidizing gas to the outer surface of oxide, where it is adsorbed.
(2) It is transported across the oxide film toward silicon.
(3) It reacts at the interface with silicon to form a new layer of SiO_2.

Each of these steps can be described as independent flux equation. The adsorption of oxidants is written as

$$F_1 = h\left(C^* - C_O\right), \tag{4.1}$$

where h, is the gas-phase transport coefficient, C^* is the equilibrium concentration of the oxidants in the surrounding gas atmosphere, and C_0 is the concentration of oxidants at the oxide surface at any given time.

It was found experimentally that wide changes in gas flow rates in the oxidation furnaces, changes in the spacing between wafers on the carrier in the furnace, and a change in wafer orientation (standing up or lying down) cause only little difference in oxidation rates. These results imply that h is very large, or that only a small difference between $C^*$and C_0 is required to provide the necessary oxidant flux.

C^* is also the solubility limit in the oxide, which is assumed to be related to the partial pressure p of the oxidant in the gas atmosphere by Henry's law

$$C^* = Hp. \quad (4.2)$$

At natural ambient pressure of 1 atm and at a temperature of 1000°C, the solubility limits are 5.2 10^{16} cm^{-3} for O_2, and 3.0 10^{19} cm^{-3} for H_2O.

The flux F_2 represents the diffusion of the oxidants through the oxide layer to the $Si-SiO_2$−interface, which can be expressed as

$$F_2 = D\frac{\partial C}{\partial x} = D\frac{C_O - C_S}{x_O}, \quad (4.3)$$

where D is the oxidant diffusivity in the oxide, C_s is the oxidant concentration at the oxide-silicon interface, and x_0 represents the oxide thickness. In this expression, it is assumed that the process is in steady state (no changing rapidly with time), and that there is no loss of oxidants when they diffuse through the oxide. Under these conditions, F_2must be constant through the oxide and hence the derivative can be replaced simply by a constant gradient.

The third part of the oxidation process is the flux of oxidants consumed by the oxidation reaction at the oxide-silicon interface given by

$$F_3 = k_s\, C_S, \quad (4.4)$$

with k_s as the surface rate constant. k_s really represents a number of processes occurring at the Si/SiO_2 interface. These may include oxidant ($O_2 \rightarrow 2O$), Si-Si bond breaking, and/or Si–O bond formation. The rate at which this reaction takes place should be proportional to the oxidant concentration at the interface C_s.

The rate at which this reaction takes place should be proportional to the oxidant concentration at the interface C_s (Fig. 4.1).

Deal and Grove assumed that in the steady state condition these three fluxes are equal, which allows expressing them as $F_2 = D\frac{\partial C}{\partial x} = D\frac{C_O - C_S}{x_O}$,

$$F_1 = F_2 = F_3 = F = \frac{C^*}{\frac{1}{k_s} + \frac{1}{h} + \frac{x_0}{D_0}}. \quad (4.5)$$

The rate of oxide growth is proportional to the flux of oxidant molecules,

$$\frac{dx_0}{dt} = \frac{F}{N} = \frac{\frac{C^*}{N}}{\frac{1}{k_s} + \frac{1}{h} + \frac{x_0}{D_0}}, \quad (4.6)$$

where N is the number of oxidant molecules incorporated per unit volume.

The differential equation can be simplified as

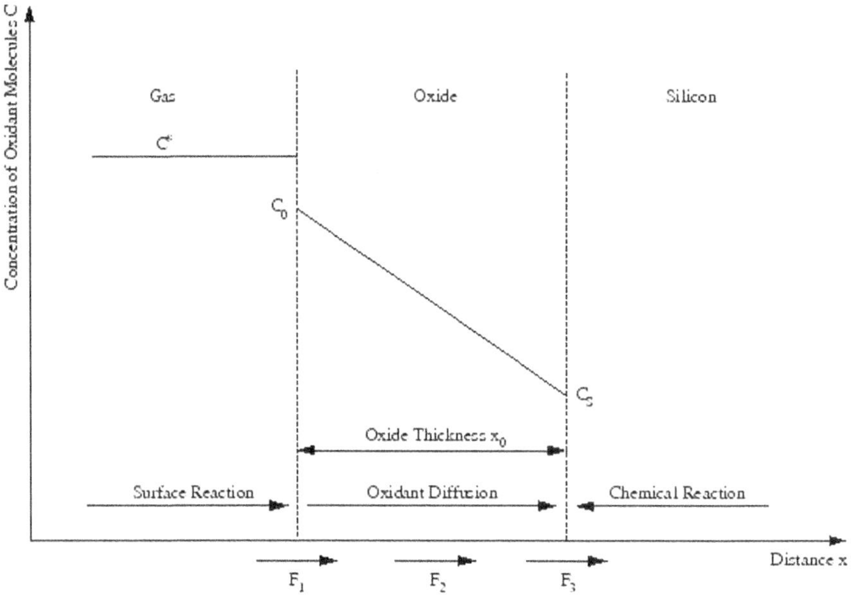

Fig. 4.1 One-dimensional model for the oxidation of silicon

$$\frac{\mathrm{d}x_0}{\mathrm{d}t} = \frac{B}{A + 2x_0}, \tag{4.7}$$

With the physically based parameters

$$A = 2D\left(\frac{1}{k_s} + \frac{1}{h}\right), \tag{4.8}$$

$$B = 2D\frac{C^*}{N}. \tag{4.9}$$

(a) *2.2 Analytical Oxidation Relationship*

In order to get an analytical relationship between oxide thickness x_0 and oxidation time *t,* the first order differential Eq. (2.6) must be solved. For this purpose in the first step (2.6) can be rewritten in the form

$$(A + 2x_0)\mathrm{d}x_0 = B\mathrm{d}t. \tag{4.10}$$

Integration of (2.9) from time 0 to t, with the assumption of an initial oxide thickness x_i at time 0, yields a quadratic equation for the oxide thickness x_0:

$$x_0^2 + Ax_0 = B(t + \tau), \tag{4.11}$$

where the parameter τ is given by

$$\tau = \frac{x_i^2 + Ax_i}{B}. \tag{4.12}$$

So τ takes into account any oxide thickness at the start of the oxidation. It can also be used to provide a better fit to the data in the anomalous thin oxide regime in dry oxidation.

At first with (2.10), the oxidation time for a specific desired oxide thickness can be estimated by

$$t = \frac{x_0^2 - x_i^2}{B} + \frac{x_0 - x_i}{B/A}. \tag{4.13}$$

(b) *2.3 Temperature dependence of B and B/A*

In order to model the corresponding growth rate for different temperatures, the values for *B* and *B/A* must change with temperature. As explained before, the oxidation rate increases with higher temperature, and so the values of *B* and *B/A* must also increase. It was found experimentally that both *B* and *B/A* are well described by Arrhenius expressions of the form

$$B = C_1 \exp\left(-\frac{E_1}{kT}\right) \tag{4.14}$$

$$\frac{B}{A} = C_2 \exp\left(-\frac{E_2}{kT}\right). \tag{4.15}$$

In these expressions, E_1 and E_2 are the activation energies associated with the physical process that *B* and *B/A* represents, and C_1 and C_2 are the pre-exponential constants. Table 4.1 lists the experimental values for the parameters needed in (4.14) and (4.15) for (111) oriented silicon at one atmosphere.

For the parabolic rate constant *B*, the activation energy E_1 is quite different for O_2 and H_2O ambient (4.15) suggests that the physical mechanism responsible for E_1 might be the oxidant diffusion through SiO_2, because *N* is a constant and C^* is not expected to increase exponentially with temperature. In fact, independent measurements of the diffusion coefficients of O_2 and H_2O in SiO_2 show that these

Table 4.1 Arrhenius parameters for *B* and *B/A* in (111) oriented silicon.

Ambient	B	B/A
Dry O_2	$C_1 = 7.72 \times 10^2\ \mu m^2/h$	$C_2 = 6.23 \times 10^6\ \mu m/h$
	$E_1 = 1.23$ eV	$E_2 = 2.00$ eV
Wet H_2O	$C_1 = 3.86 \times 10^2\ \mu m^2/h$	$C_2 = 1.63 \times 10^8\ \mu m/h$
	$E_1 = 0.78$ eV	$E_2 = 2.05$ eV

parameters vary with temperature in the same way as (2.18) and with E_l values close to those shown in Table 2.1. The clear implication is that B in the linear parabolic model really represents the oxidant diffusion process.

(i) LPCVD (ii) APCVD (iii) PECVD

4.3 Chemical Vapor Deposition (CVD)

In this process, heat and mass are transferred through convection mechanism and chemical reaction takes place on the surface. CVD process results in better quality and process control. Transport gas diffuses on the full substrate and process is carried on lower gaseous pressure. CVD process depends on various parameters such as:

- Temperature.
- Pressure of the carrier gas.
- Velocity of the gas.
- Distance in the direction of gas transport (Table 4.2).

Various parameters such as RF energy, pressure, gas ratio, and stresses are mainly controlled with the temperature.

$$\begin{aligned} &SiCl_4 + 2H_2 \text{ (gas)} && -> Si\text{(solid)} + 4HCI[1250\,^{\circ}C] \\ &SiH_4 \text{ (gas)} && -> Si\text{(solid)} + 2H_2\text{(gas)}\left[1000\,^{\circ}C,\ \text{temperature dissolution}\right] \end{aligned} \tag{4.16}$$

1. Poly silicon

$$SiH_4 \text{ (gas)} \quad -> Si \text{ (solid)} + 2H_2\text{(gas)}\left[600 - 700\,^{\circ}C\right] \tag{4.17}$$

2. Silicon dioxide (silicon dioxide) (Fig. 4.2)

$$\begin{aligned} &SiH_4\text{(gas)} + 2O_2 \quad -> Si\text{(solid)} + 2H_2O\text{(gas)}\left[450\,^{\circ}C\right] \\ &SiH_4 + O_2 - SiO_2 + 2H_2 \\ &SiCl_2H_2 + 2H_2O - SiO_2 + 2H_2 + 2HCl(900\,^{\circ}C) \end{aligned} \tag{4.18}$$

Table 4.2 General characteristics of PECVD layer

Deposited layer	Characteristics	Applications
Silicon oxide	Low thermal conductivity Low mean stress Selectivity	Sacrificial layer Wafer bonding Electrical and thermal isolation
Silicon nitride	High quality factor Low elastic force	Insulating layer

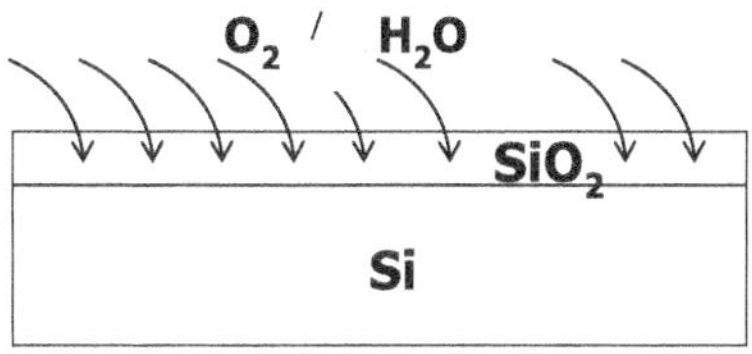

Fig. 4.2 Silicon dioxide process

Table 4.3 Various deposition techniques

Chemical vapor deposition (CVD)	Physical vapor deposition (PVD)	Wet deposition technique
LPVD, PECVD MOCVD ALDC atomic layer	Sputtering Vaporization Atomic beam Laser	Electrodeposition SOI Gel Solvent

3. **Silicon Nitride**

$$
\begin{aligned}
&3SiH_4 + 4NH_3 - \rightarrow Si_3N_4 + 12H_2 \ (700 - 900\,^\circ C)\\
&3SiCl_4 + 4NH_3 \rightarrow Si_3N_4 + 12HCl \ (800C)\\
&3SiH2Cl_2 + 4NH_3 \rightarrow Si_3N_4(solid) + 6HCl + 6H_2(gas)(650 - 750\,^\circ C)
\end{aligned}
\tag{4.19}
$$

Isotropic etching (HNA)

$$
\begin{aligned}
&Si + 4HNO_3 \rightarrow SiO_2 + 2H_2O + 4NO_2\\
&SiO_2 + 6HF \rightarrow H_2SiF_6 + 2H_2O
\end{aligned}
$$

4. **Thermal Oxidation**

$$
Si + 2H_2O \rightarrow SiO_2 + 2H_2(\textbf{liquid oxidation})900 - 1200\,^\circ C \tag{4.20}
$$

5. (Table 4.3)

$$
Si + O_2 \rightarrow SiO_2 \ (\text{dry oxidation}) \tag{4.21}
$$

4.4 Sputtering

It is the physical reaction in which ions are accelerated toward the target due to voltage gradient. The atoms are deposited in the substrate due to momentum transfer. Mostly argon gas is employed for the ionization and DC/RF voltages are employed.

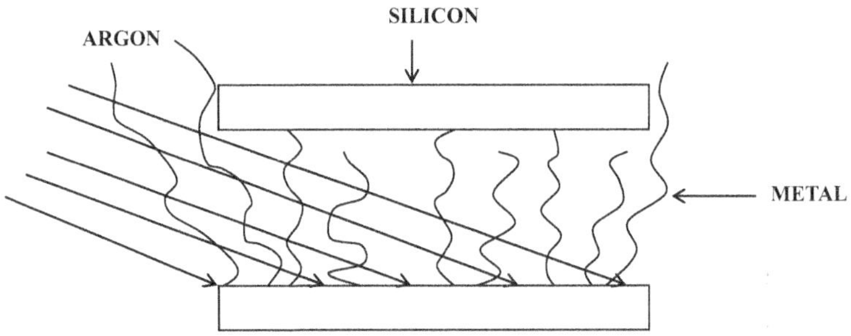

Fig. 4.3 Metal sputtering process

Deposition rate depends on the material, RF energy, chamber pressure, and electrodes distance. Deposited layer is mostly amorphous and can be crystallized by annealing (Fig. 4.3).

The main process steps in the surface micromachined process for the cantilever are:

1. The thick layer generation of silicon oxide.
2. Deposition of photo-resist layer over the oxide surface.
3. Patterning (to segregate the cantilever geometry).
4. Etching of the oxide.
5. Cavity structure in the silicon wafer using the anisotropic etching.

Various materials are combined in the surface micromachining process such as polysilicon, silicon oxide, polyimide, aluminum, silicon nitride.

4.5 Vaporization

Metallic films are deposited using this process. The material is kept in the crucible and vacuum is created in the chamber. The material is vaporized with high current or electron beam. The deposition of material is dependent upon the distance and time.

4.6 Lithography

The transferring process of the feature from the mask on to the optical sensitive material is termed as lithography. The UV rays are employed in this process is termed as photolithography. The photo-resist is categorized into two types:- positive and negative (Table 4.4)

Table 4.4 Photo-resist types

Parameters	Positive	Negative
Commercial example	PR120	SU8, kodak 747
Adhesion with silicon surface	Better	Good
Step coverage	Better	Good
Resistance to the plasma etching	Very effective	Not effective

Main process steps are:

(i) Resist deposition.
(ii) Baking.
(iii) UV exposure.
(iv) Patterning.
(v) Baking.
(vi) Etching.
(vii) Removal of photo-resist.

MEMS circuits are based on the capacitive and resistance sensing mechanism, so the stress in the thin films effect the overall behavior and reliability. Silicon nitride is having low thermal conductivity and its mechanical characteristics are stable at the cryogenic temperatures. The thermal conductivity of silicon oxide is low and this layer is used for the electrical isolation.

4.7 Etching

This process may be of two types, namely: isotropic and anisotropic.

Unwanted material is removed in the etching process. Bulk micromachining process the other material is deposited as hard material and silicon is taken out. This is the chemical process where solid state is not achieved. The etching is carried out in liquid or gas phase. The etching process is characterized by etching rate, etching selectivity, etching similarity and divided into wet and dry process. In the case of dry phase, mutual interaction between chemical and physical phenomena is involved whereas in the wet techniques based on the chemical reaction where etching solution and substrate interact to remove the undesired material (Table 4.5).

Dry Etching Equation:

$$\begin{aligned} CF_4 &\rightarrow CF_3 + F^* \\ Si + 4F^* &\rightarrow SiF_4 \end{aligned} \tag{4.22}$$

Main steps in the etching are:

1. Insertion of hole in the substrate.
2. Positive charge associated with the negative element.

Table 4.5 Wet etching chemistry for traditionally used materials in MEMS

S/N	Material	Etching chemistry
1.	Gold(Au)	Aqua-rijia ($HCl:HNO_3$)
2.	Platinum(Pt)	Acqueo-riza@ 85 °C
3.	Aluminum (Al)	$H_3PO_4 : H_2O : CH_3COOH; HNO_3$
4.	Chromium/Nickel (Cr/Ni)	$HCl : HNO_3$
5.	Silicon nitride	$H_3PO_4$150 °C
6.	Silver(Ag)	$NH_4OH : H_2O_2$
7.	Titanium(Ti)	$H_2O : HF : HNO_3$
8.	Zinc(Zn)	HCl/EKC
9.	Zinc oxide	$CH_3COOH : HPO_4 : H_2O$

3. Mutual reaction.
4. Dissolution of remaining products in the etching solution.

Isotropic and anisotropic etching is possible in the crystalline material such as single crystal silicon and etching depends on the choice of the chemical.

4.7.1 Isotropic Etching

In this process, the etching rate is dependent on the crystal direction and etching is same in all directions. The etching rate is dependent on the temperature and impulse. This process control of etching in the vertical and horizontal direction is difficult. HNA is the main chemical for isotropic etching of the silicon.

$$Si + 4HNO_3 \rightarrow \quad SiO_2 + 2H_2O + 4NO_{2.} \tag{4.23}$$

This process is used for the thinning of the wafer. Pressure sensor fabrication consists of bulk and surface micromachining processes. Silicon is etched using various techniques which are detailed as below (Tables 4.6 and 4.7):

4.7.2 Anisotropic Etching

Etching is dependent on the plane orientation, and in case of silicon semiconductor material, some specific planes are etched much faster compared to the other planes. The dissolution of some specific plane depends upon the packing density and in case of silicon wafer <111> plane is having lesser etch-rate compared to other planes. The angle intended between <100> and <111> is $54.74^0$and in case of <110> and <111> is 90.0^0. Mainly, KOH, TMAH, EDP, and NaOH are employed for etching

Table 4.6 Various types of silicon etching chemistry

S. No.	Parameters	$HF:HNO_3: CH_3COOH$	KOH	EDP	TMAH	DRIE
1.	Type	wet	wet	wet	wet	wet
2.	Anisotropic	No	Yes	Yes	Yes	Yes
3.	Rate(min/micron)	1-20	0.5-2	0-7.5	0.5-1.5	1-3
4.	Selectivity <111/100>	No	100:1	35:1	50:1	No
5.	Nitride etching	Lower	<1	0.1	<0.1	200
6.	Oxide etching(micron/min)	30-10	10	0.2	<0.1	10
7.	End of etching P++	No	Yes	Yes	Yes	No

Table 4.7 Generic etchant and its composition

Name of the material	Etchant details
Silicon	KOH/TMAH/ or 3 ml HF + 5 ml HNO_3+ 3 ml CH_3COOH
Silicon Oxide	28 ml HF+ 170 ml H_2O+ 113 gm NH_4F
	15 ml HF+ 10 ml HNO_3+ 300 ml H_2O
$Si3N_4$	Buffered HF
	H_3PO_4
Al	1 ml HNO_3+ 4 ml CH_3COOH + 4ml H_3PO_4+1 ml H_2O
Au	4 gm KI+ 1 gm I_2+ 40 ml H_2O
TiW	H_2O_2

the silicon wafer. The orientation of the mask with respect to the wafer is important and mask is kept collinear with <110>; otherwise, it may lead to lateral undercut, V-groove, or pyramid pit. Etch stop technique is employed to close the etching.

4.8 Micromachined Implementation

Micromachining is the basic operation carried out in MEMS and this process, particularly in RF, leads to higher frequency operations in transmission lines due to minimal surface-wave losses. The process leads to opening from the backside which can be filled by using CMOS compatible material such as polyimide. This facilitates the lowering of permittivity as well as ease of backside metallization without any extra processing steps. Further to ease out implementation of microstrip realization, the machining portion is filled with polyimide which is further cured. The added advantage of polyimide is the CMOS compatibility. This assembly is metalized from backside without any extra processing steps to achieve microstrip configuration.

Bulk micromachining is basic operation in MEMS resulting in the removal of the bulk substrate either wet or dry etching technique. This process results in various

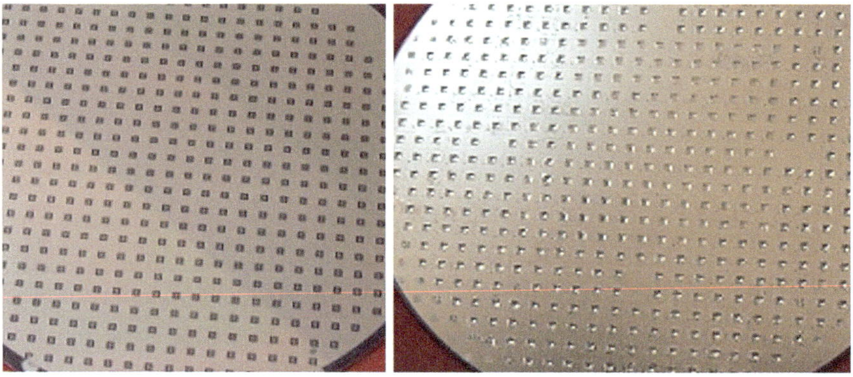

(a) (b)

Fig. 4.4 Interdigital structure on the wafer. **a** Front side **b** Micromachined back side

advantages of planar transmission line compared to the realization of transmission line on soft substrates. The basic advantages are:

- Minimal surface-wave propagation.
- High-frequency operation (Fig. 4.4).

Another mask is used for the backside processing for which micromachining is employed to create cavity at the desired location.

Planned micromachining steps are the combination of DRIE and KOH etching to have controlled and reliable machined process with mature technology. Limitation of total DRIE or KOH machining is the holding and patterning of wafer after DRIE, whereas KOH attacks other layers vigorously, thus spoiling full circuit. Also concept of oxide and nitride is taken as oxide itself acts as a hard mask for DRIE and nitride is used as a masking layer for KOH. Bulk micromachining is carried out by employing wet or dry techniques for isotropic and anisotropic etching. Wet anisotropic etching carried out using KOH can be represented as (Fig. 4.5)

$$\begin{aligned} &\mathrm{Si} + 2\mathrm{OH}^- = \mathrm{Si(OH)}_2^{2+} + 4\mathrm{e}^- \\ &4\mathrm{H_2O} + 4\mathrm{e}^- = 4\mathrm{OH}^- + \mathrm{H_2} \\ &\mathrm{Si(OH)}_2^{2+} + 4\mathrm{OH}^- = \mathrm{SiO_2(OH)}_2^{2-} + 2\mathrm{H_2O} \\ &\mathrm{Si} + 2\mathrm{OH}^- + 2\mathrm{H_2O} = \mathrm{SiO_2(OH)}_2^{2-} + \mathrm{H_2}. \end{aligned} \tag{4.24}$$

Mathematically depth can be represented as

$$w_0 = w_m - 2\ \cot\ \mathrm{an}(54.74°)z. \tag{4.25}$$

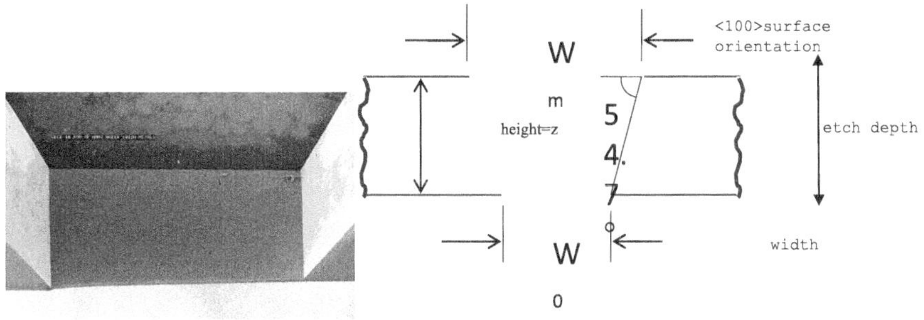

Fig. 4.5 Bulk micromachining and measurement of depth

The above process is implemented in the interdigital topology. The critical process parameter is the masking layer thickness to protect the undesired etching. In the present process, thick oxide layer is used for masking and customized jig to protect the front side is employed for etching the wafer.

4.9 Case Study: MEMS-Based Acoustic Sensor and Process Challenges Associated with It

Micromachined acoustic devices are having applications in various domain such as hearing-aid, cellular phone, micropersonal digital assistant, aerospace for measuring sound pressure and earphone [1]. Sensor based on piezo-electric layer is having more robust process, and further, CMOS-MEMS sensor has the advantage of on-chip electronics. The sensor having piezo-electric layer is sandwiched between two thin film layers with backside having closely spaced tunnel and cavity. Micromachined-based sensor is based on the piezo-electric layer which converts the sound pressure into electrical energy. Preferably zinc oxide (ZnO) layer is employed as piezo-electric layer but other piezo-electric materials such as lead zirconate titanate (PZT), PVDF, $PbTiO_3$, $LiNbO_3$ crystals, and AlN can be used but compatibility and ease of processing are the important considerations. Zinc oxide is having excellent piezo-electric and dielectric properties so piezo-electric film and sensors based on it are prominently employed for the realization of the acoustic sensor [2], gas sensors, microphones, ultrasonic transducers, FBAR device which finds applications in space, defense, and commercial domain [1].

In this section, diaphragm-based sensor with ZnO as piezo-electric layer and backside incorporation of microtunnel for the compensation of charge with sound pressure during diaphragm vibration is presented. The cavity is etched to achieve desired diaphragm thickness whereas tunnel facilitates pressure equalization of the diaphragm. Thus, depth of tunnel and cavity is varied so as to achieve the desired performance [3]. Various techniques such as DRIE, surface chemical treatment, corner compensation are employed to achieve differential etching but are having limitations either due to uncontrolled etching or due to convex corner undercutting phenomena [4]. Lateral etching observed in silicon bulk micromachining using wet etching techniques while etching of <100> surface and various compensation techniques proposed to overcome these phenomena such as mask having square/rectangular/conical shapes, <110> oriented beams but compromise structure compactness and surface roughness [5]. Dry etching for cavity and tunnel can be carried out using DRIE process but the process become complex and yield different cavity geometry which can change the resonance frequency. Alternatively, wet etching employing KOH and TMAH can be employed but the process yields convex corner undercutting phenomena.

Present developments focused on the micromachined process for which various trials are conducted to differential etch cavity and tunnel. Differential etching techniques are employed to create cavity of ~3mm × 3mm with ~500 um depth and tunnel width of 35 um with 100 um depth. Further process trials, thin film and sensor characterization are also presented.

4.9.1 Acoustic Sensor Process

SOI wafer is employed as the base layer having the added advantage of buried silicon dioxide layer which acts as an etch stop for KOH solution and also having good selectivity over SiO_2. This machining process also ensures uniformity of diaphragm thickness. The overall cross section is shown in Fig. 4.6.

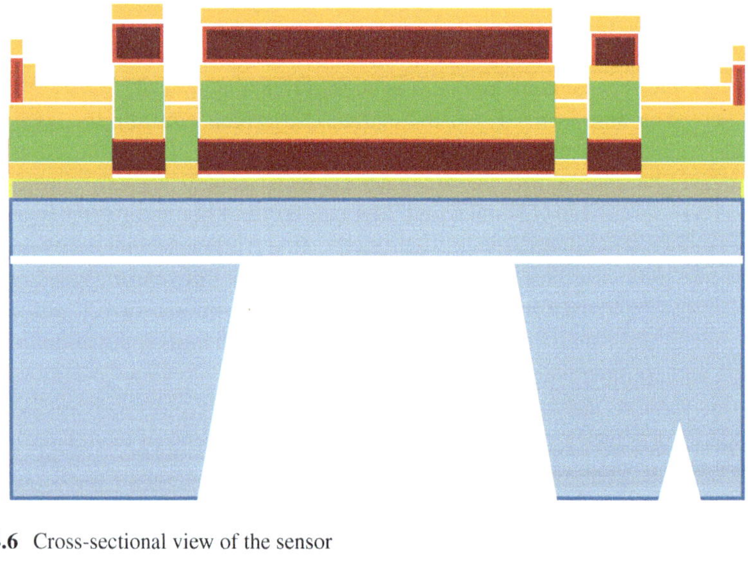

Fig. 4.6 Cross-sectional view of the sensor

Table 4.8 Typical etch-rate of masking layer

Material	Etchant	Typical etch-rate (Å/s)
PECVD oxide	BOE	> 70
LPCVD oxide	BOE	< 15
Grown oxide	Dry etching	> 85
Deposited oxide	Dry etching	> 100

Stress and etch-rates of various layers are analyzed before finalizing the process (Table 4.8). The piezo-electric layer of ZnO deposition is carried out on the front side and the flow chart depicting the back end process is shown in Fig. 4.7. The lateral stress on the zinc oxide layer generates charge which is converted into voltage.

The process starts with selection of the base material which is taken as SOI where oxide and nitride layers are deposited (Fig. 4.7). The process employ low pressure based chemical vapor deposition of oxide followed by lithography. Polyimide is taken as photo-resist which is a negative photo-resist material. Dry etching for back-side removal of oxide and nitride followed by thick oxide deposition using PECVD process carried out which is having low mean tensile stress (Fig. 4.7).

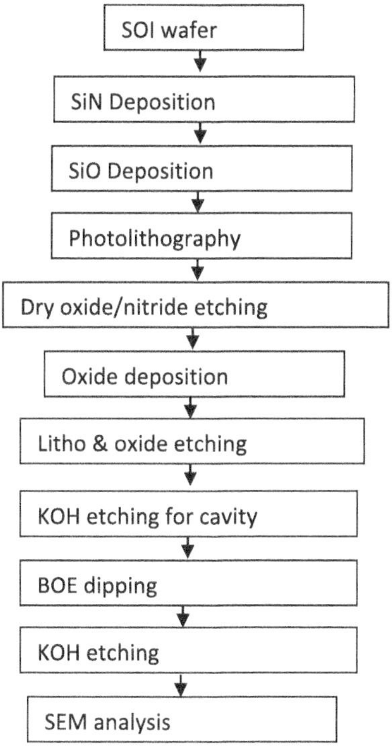

Fig. 4.7 Back side process steps

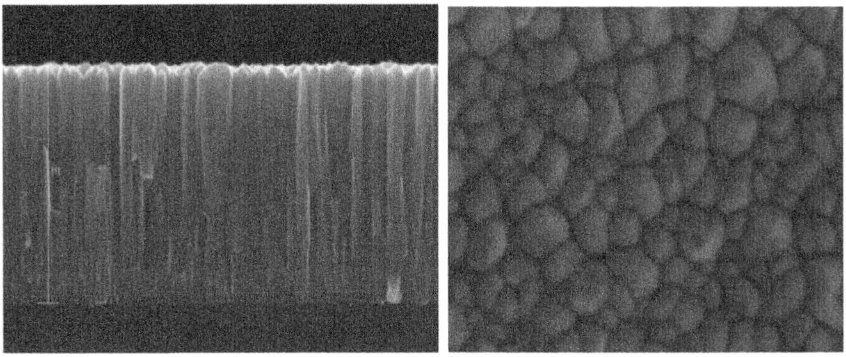

(a) SEM image of deposited layer (b) AFM image of the layer

Fig. 4.8 SEM and AFM image of the deposited ZnO layer

Front side ZnO layer is deposited using sputtering [6] and its characterization employing various techniques such as XRD, AFM, SEM, STM are carried out (Fig. 4.8).

Thickness uniformity and controlled grain size are important to achieve desired performance.

4.9.2 Various Process Trials in Acoustic Sensor

The present requirement of micromachining is to etch tunnel and cavity with unequal depth as per design requirements. The critical aspects associated with the differential etching is the prolong exposure of the wafer which lead to exposure of various other planes such as <411>, <311>. This process eventually leads to convex corner undercutting. The cavity and tunnel as required on the backside of the die are shown in Fig. 4.9.

The tunnel and cavity etching carried out using the compensation structures leads to lateral undercut in the present case of deep etch cavity (Fig. 4.10)

Another trial employing DRIE-based process yields same depth etching so modified process having the combination of dry and wet is tried. DRIE-based process yields same depth and the shape of the cavity also changed to rectangular instead of trapezoidal [7]. Further combination of wet and dry etching technique yields the desired structure but results in the variation in the cavity shape which impacts the cut off frequency (Fig. 4.11). Also tight etch control in these structures is difficult to achieve. Modified process is implemented to achieve the desired depth is carried out in phased manner. First phase oxide is removed from the cavity region and etching the cavity until certain depth using the corner compensated mask (Fig. 4.12). Etch

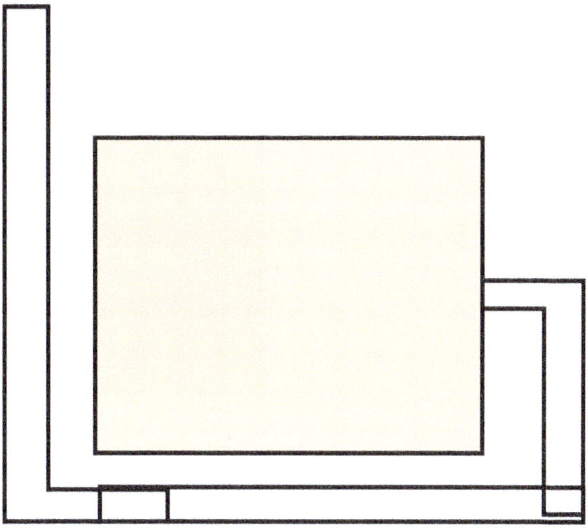

Fig. 4.9 Schematic of backside topology for acoustic sensor

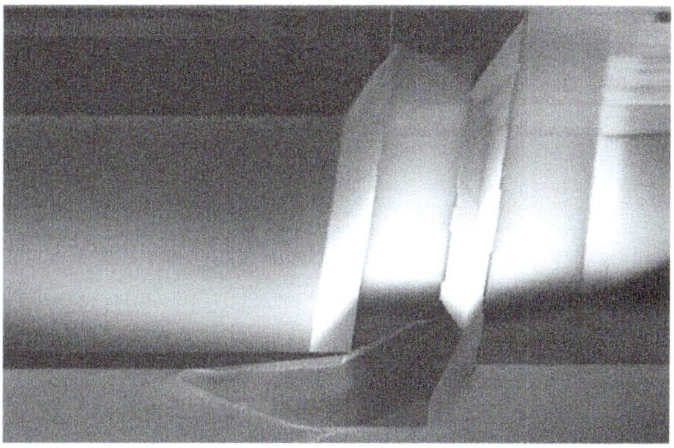

Fig. 4.10 Convex undercutting in the development of micro-tunnel for acoustic sensor

depth calculation can be carried out by deducing the etch-rate and in situ measurement technique. In the second phase, oxide from the tunnel region is also removed and simultaneous etching of both cavity and tunnel carried out. This process yields desired cavity depth of 500 ± 25 um and tunnel depth of 35 ± 10 um.

4.9.3 Measurement Results

Main parameters in sensor measurement are capacitance which affects the voltage and charge sensitivity. Further loss tangent value provides insight of the layers quality as well as affects the linearity of the device. The measurement is carried out after assembly of the devices. LCR meter is employed to measure the capacitance value

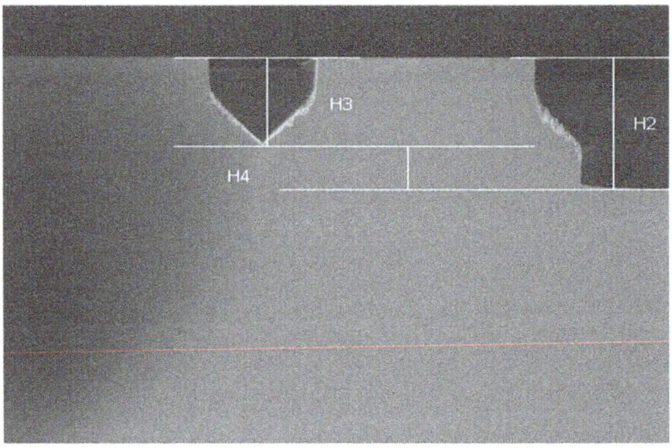

Fig. 4.11 Combination of dry etching with aqueous KOH etching

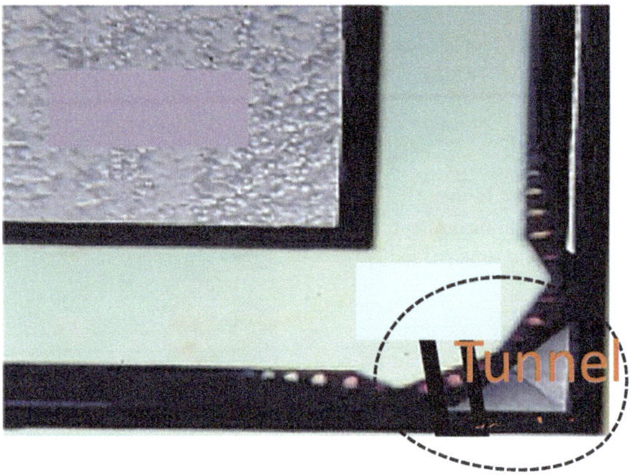

Fig. 4.12 Final structure with modified process

Controlled die to die variation (Table 4.9) shows the repeatability of the process, whereas stable capacitance and low dissipation value indicates least deviation resulting in better voltage and charge sensitivity.

Table 4.9 Characterization of acoustic dies

Die no	Capacitance (pF)	Dissipation factor
1	110	0.05
2	100	0.04
3	105	0.02
4	115	0.02
5	105	0.03
6	105	0.03
7	106	0.05
8	104	0.04

Table 4.10 Overview of various techniques

Technique	Remarks
Convex compensation technique	Ineffective in case of narrow channel and limited gap
Surfactant	Only increases the etching depth
DRIE-based process	Single mask micromachining not possible with high differential ratio
Differential etching(KOH)	Two mask process

This section details the mitigation techniques for overcoming convex undercutting phenomena while carrying out differential etching. The present process for differential etching of the sensor topology is fully compatible with the CMOS process. Narrow tunnel and deep cavity is simultaneously etched by combination of two phase etching with corner compensation technique yielding desired geometry. Also capacitance and dissipation factor measurement of the sensor capacitance shows the stability of the process. The process is fully compatible with CMOS paving way for electronics integration on the single chip (Table 4.10).

4.10 Case Study: MEMS-Based Micromachined Antenna Array

Fabrication of the antenna structure starts with standard 6" high resistive silicon wafer of 675 $\pm$ 20 μm thickness and resistivity of the order of 8 kΩ-cm. After giving the conventional chemical cleaning treatment, 500 Å base oxide is grown on the wafer, which acts as buffer layer for the subsequent RF circuits printed on this. Aluminum (1 μm) is sputtered and patterned subsequently on the front side of the wafer to realize metallic antenna patches with feed network. A thick passivation or protective layer is coated thereafter on the front side, which is followed by thinning of wafer up to 440 μm with KOH solution. This thin wafer is patterned and etched in KOH solution from the backside to form cavities with 50 $\pm$10 μm membrane thickness underneath patches. Finally, the front side passivation layer is removed in suitable solution. Fig. 4.13 depicts the top and bottom views of fabricated micromachined prototype antenna array.

For characterizing the circuit, the fabricated structure is diced with standard dicing tool and then it is assembled on aluminum jig (working as microstrip ground plane) with RF connector (2.92 mm), shown in Fig. 4.13. With the present in-house facility, the return loss characteristic has been measured using R&S make Vector Network Analyzer ZVA-40. And, the far-field radiation pattern measurement was carried out in CATF facility of SAC, Ahmedabad. Simulated 3D radiation pattern is shown in Fig. 4.14. It shows that there is dip in the gain value at the boresight direction. Peak gain of the antenna array is coming around 12.8 dBi.

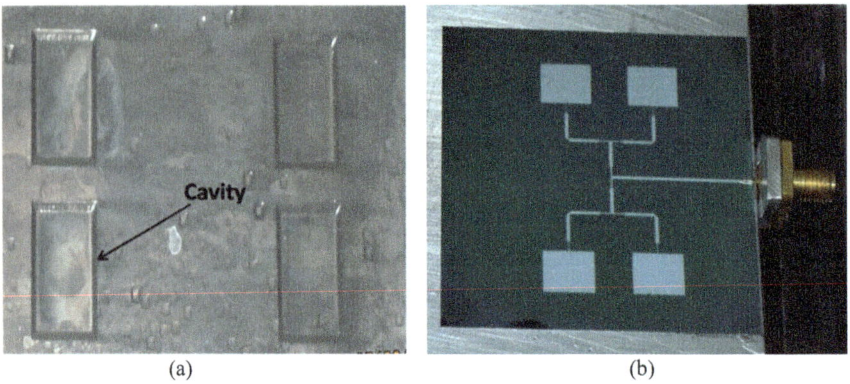

Fig. 4.13 **a** Bottom view of the micromachined antenna array. **b** Test fixture with antenna array mounted

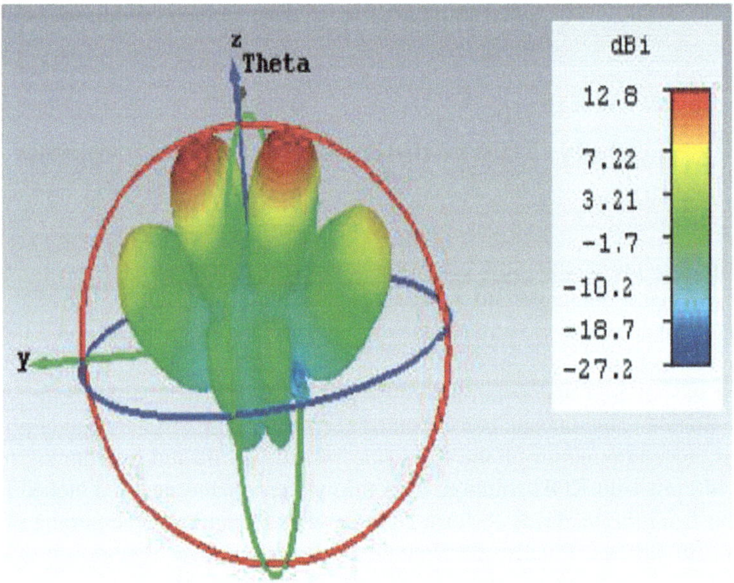

Fig. 4.14 Simulated far-field radiation pattern (3D) of the prototype antenna

This kind of antenna can find wide applications in wireless and RADAR. Specially, in radar communication, while a jamming signal is essential, difference pattern of the antenna is generated. The difference pattern consists of a null in the boresight direction with two major lobes adjacent to null.

4.11 Case Study: MEMS-Based Cantilever Realization

Main steps in the fabrication of the cantilever are as below

- Thick layer of silicon oxide.
- Photo-resist over oxide.
- Identification of etching area by lithography.
- Etching of unwanted oxide.
- Cavity structure due to anisotropic etching of silicon wafer.

Various materials combination are taken in surface micromachining such as polysilicon, silicon dioxide, polyimide, aluminum, silicon nitride, etc. (Fig. 4.15).

Various cantilever structures are formed with these techniques. The realized microstructure are realized by deposition and etching to realize various structural layers (Fig. 4.16).

$$\begin{aligned} 2HF + H_2O &\leftrightarrow H_3O^+ + HF_2^- \\ SiO_2 + 2H_3O^+ + HF_2^- &\leftrightarrow SiF_4 + 4H_2O \end{aligned} \tag{4.26}$$

The empty place is filled with the dry nitrogen so that humidity and changes in gas resonance frequency due to gas density can be made effective. Dry technique which basically employs deep reactive ion etching (DRIE) results in anisotropic etching. Main processes in MEMS can be segregated as depicted in Fig. 4.17.

Fig. 4.15 Surface micromachined structure

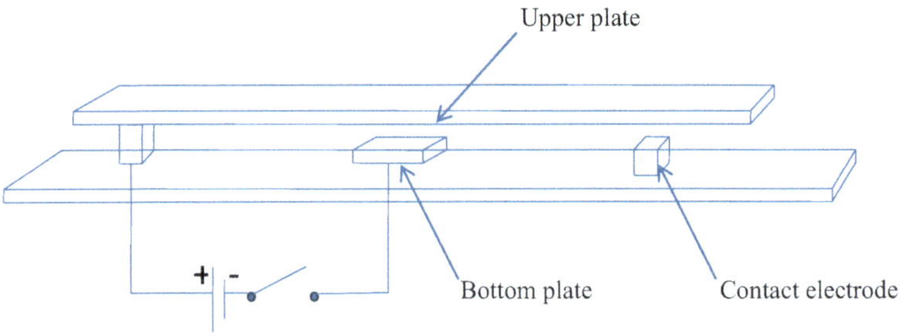

Fig. 4.16 Schematic of cantilever beam

Fig. 4.17 Flow chart of generic MEMS fabrication

MEMS Design
↓
Deposition/Diffusion
↓
Etching (Dry/wet)
↓
Ion implantation
↓
Thin film deposition
(Bulk/surface micromachining)
↓
Analysis
↓
Testing
↓
Packaging
↓
Reliability Testing

4.12 Summarization

This chapter highlights various microfabrication processes involved in the realization of MEMS devices. There are several common fabrication process steps as that of standard CMOS process. The unique features of MEMS technology are lying in the four major processes, namely wafer bonding, bulk micromachining, surface micromachining and electroplating. Free-hanging structures can be realized with these processes, which is hardly possible to achieve in standard microfabrication process. Three case studies have also been covered in this chapter to get as flavor of practical implementation. Process intricacies have also been emphasized in this chapter.

Questions

1. What are the different sensing mechanisms available in MEMs technology? Compare them.
2. What is piezo-electric effect? Name one sensor, where it is used.
3. What are the usages of acoustic sensor?
4. How micromachining can be useful for antenna engineering?
5. What are the challenges associated with surface micromachining method?
6. What is bulk micromachining method? How is it achieved?
7. Name two dry and wet etchants of silicon substrate.
8. How the anisotropic etching is done for silicon?
9. Name few etch stop techniques for silicon.
10. How the non-uniformity of etching is measured?
11. How do you calculate the etch-rate of some etchant?
12. Why KOH is most preferred wet etchant for silicon wafer?
13. Name some substrates other than silicon where micromachining can be implemented.
14. How do you measure the surface profile of the etched micromachined cavity?
15. How the surface roughness is measured? Tell the working principle of the machine.

References

1. Ananthsuresh GK et al (2005) Micro and smart systems. Wiley India
2. Prasad M, Shukla V, Khanna VK (2013) Design and fabrication of Si diaphgram ZnO piezo electric film based MEMS acoustic sensor using SOI wafers. IEEE Trans. Semiconductor Manuf. **26**(2):233-236
3. Arora A, Arora A, Dwivedi VK, George PJ, Srrnivas K, Gupta V (2008) Zinc oxide thin film based MEMS acoustic sensor eith tunnel compensation. Sensors Actuators, A141:256–261
4. Bao M, Burrer C, Esteve J, Bansells J, Marco S (1993)Etching front control of <110> strips for corner compensation. Sens Actuators A. 37–38:727–732
5. Puers B, Sansen W (1990) Compensation structures for convex corner micromachining in (100) silicon. Sens Actuators, A21-A23:1036–1042
6. Singh K, Nirmal AV et al (2019) Process optimization and thin film characterization of ZnO layer for sensor applications. IJEEI J. 2
7. Nitesh et al (2017) Resistance analysis of acoustic sensor vent tunnel. INAC-03

Chapter 5
Thin Film Stress

Generally, a substrate and thin film over it have different thermal expansion coefficients, so stress is produced during temperature changes occurring in deposition and annealing. In microelectronic devices, the stress can be a major cause for yield and reliability problems. Even it can alter the actual device properties (for optoelectronic devices). A piece of solid is said to be under stressed while displaced from their equilibrium positions by a force. The displacement is governed by the inter-atomic potentials.

Stresses in the thin film are primarily tensile or compressive. These stresses are either due to the metal layer or dielectric layers. Dielectric films deposition based on thin film technology using Plasma Enhanced Chemical Vapor Deposition (PECVD) processes are commonly employed in microelectronics fabrication. Oxide and Nitride are the key dielectric layers in any microelectronics process. Deposition parameters play an important role in such deposition and can drastically affect overall performance of the device. Stresses in these films can affect the electrical and optical characteristics as well as the integrity and reliability of the devices. The stresses induced due to thickness variation and annealing temperature also to be characterized for better design and reliable circuit operations.

5.1 Introduction

Typically the microelectronic device/micro-electro-mechanical system (MEMS) product fabrication starts with the deposition of oxide and nitride layers over the silicon. Usually, PECVD and LPCVD are the processes to deposit thin dielectric layers. Deposited layers play an important role in annihilation of substrate defects which in turn introduce stresses which is dependent on thicknesses. Also temperature plays an important role in deposition and in creation or annihilation of stress. For MEMS devices based on capacitive/resistive sensing mechanism, the stress

K. Singh and A. Karmakar, *Micro Electro Mechanical System (MEMS) Technology for Multifaceted Applications*, https://doi.org/10.1007/978-981-96-6162-6_5

phenomenon drastically affect the overall performance and hence the reliability of the MEMS sensors.

The thin films must be mechanically stable. The process parameters (temperature, pressure, gas chemistry, power density, frequency of the RF excitation, etc.) influence greatly the stress profile of the deposited films. Characterizing the stress-state is very essential for micro-devices.

Different types of materials such as SiO_2 or Si_xN_y are being used in MEMS as structural material, especially in the fabrication of suspended membranes for pressure or micro-heaters-based sensors. Residual stress can be compressive, which makes the film expand parallel to the surface, or tensile, which makes the film shrink [1]. The profile as well as the robustness of a thin dielectric membrane is affected by the residual stress. A compressive membrane may buckle, avoiding its use to support sensors. On the other hand, too tensile membrane can break in presence of pressure or high temperature gradients. It is therefore essential to be able to measure and control the stress separately in each layer composing structure in order to design reliable multi-layers.

Donald [1] mentioned about the measurement of stresses on the globally deposited wafers by measuring the curvature/bow in the wafer. Guckel et al. [2, 3] developed some test structures to monitor the stress during deposition process.

While considering the reliability issue of the micro-devices, one of the important concern is the development of 'residual stress' in it. It may be a major cause of device failure. It is developed due to differences in the coefficient of thermal expansion of various thin films and substrate material with which the device is made of growing mechanism of the thin film doesn't allow the atoms to remain in the least energy state or in other words in the sable state. Either the atoms go further apart than they would be in the annealed state (tensile stress) or they will come closure together (compressive stress).

Various studies [4] conducted show the behavior can be altered with annealing temperature. Annealing temperature has major effect for certain thickness and can be optimally selected to nullify stresses in the film. PECVD-based thin film deposition employing both TEOS as well as silane-based chemistry is extensively dealt in this chapter. Zinc oxide is another important thin layer. Deposition of this metal-oxide thin film is an important process in the realization of MEMS-sensor. Its piezo-electric and optoelectronic characteristics are of unique type. Performance of the final sensor depends upon several metrics of sensor, such as orientation of the crystal used, morphology of the substrate, and dielectric constant of the material used. Various other parameters, like grain size of the crystal, reflectance of the material, band-gap of the material used, etc., are also critical in the development of final sensor.

Here, below in Fig. 5.1 the direction and sign of applied force by convention is shown:

In Fig. 5.2, a piece of thin solid film of dimension $l \times w \times t$ is depicted. If a force F is applied to the area A $(= w.t)$ to stretch the film length l by Δl, we have

$$\frac{F}{A} = Y.\frac{\Delta l}{l} \text{ or } \sigma = Y.\varepsilon \tag{5.1}$$

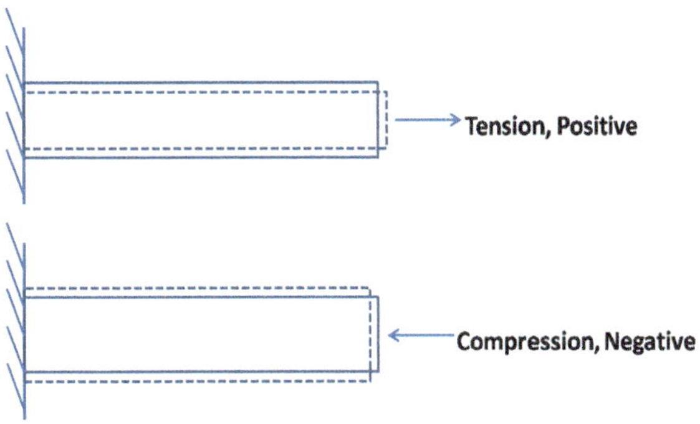

Fig. 5.1 Direction of force to make understand of tensile or compressive nature

Fig. 5.2 A piece of thin film of dimension (l × w × t) under tension

where $\sigma = F/A$ and $\varepsilon = \Delta l/l$ are stress and strain, respectively and Y is Young's modulus. This is well known Hooke's Law. In addition,

$$\frac{\Delta t}{At} = \frac{\Delta w}{w} = -v.\frac{\Delta l}{l} \tag{5.2}$$

where v is the Poisson's ratio. It is generally a positive number for almost all materials and is less than 0.5. The negative sign in Eq. (5.2) signifies that while we stretch l, both 'w' and 't' shrink.

In Fig. 5.3, we can observe how the deformation and fracture occur in stressed thin film.

The residual stresses are stored energy than can be released in subsequent processing steps or with time in storage or in service. Total stress inside the thin film is a function of film thickness, elastic modulus, and morphology of the film and density of it. Stress can easily reach the level where plastic deformation materials (Cr, W, or metal oxides), very high stresses can be attained with a great deal, whereas for low modulus material (Au, Ag, etc.), plastic deformation will relieve the stresses before high stress levels are achieved. The stresses within the film may not be uniform throughout the thickness always. And, because of this the film will curl when separated from the substrate.

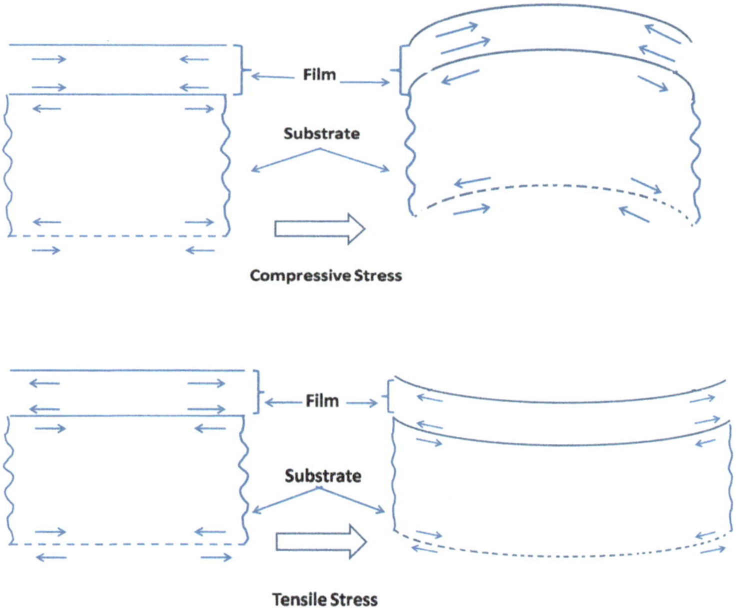

Fig. 5.3 Deformation and fracture in stressed films

Fig. 5.4 Stress measurement technique

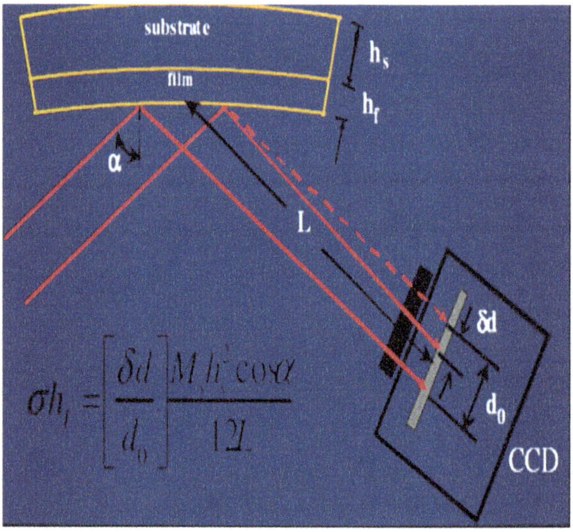

5.2 Process for Oxide/nitride Deposition

All the SiO_2 and SiN_x films with varying thickness are deposited using PECVD system. The PECVD reactor is a of a conventional parallel plate design and uses a high frequency (13.56 MHz) power source to generate the plasma. For the stress experiments, 2 types of films, SiO_2 and SiN_x were deposited. The doped SiO_2 and SiN was deposited with silane chemistry. The oxide is deposited using SiH_4, N_2O, and N_2 at a temperature of 400 °C. Similarly the nitride was deposited using silane chemistry. The processing conditions were kept as following:

Pressure	2.5 Torr
Temperature	400 °C
SiH_4/NH_3	1:6.5
RF power (W)	60

5.2.1 *Stress Measurement*

Stress measurement of each layer is carried out using stress measurement tool [K-space]. Stress in thin films induces curvature in the substrate. The system measures localized curvature by monitoring the deflection of parallel beams of light. Curvature mapping of the entire sample surface is performed by 2D iteration via high precision, servo motor controlled x–y stage. Stress is calculated using Stoney's formula as follows:

$$\sigma_{film} = \frac{1}{6R}\frac{E_s t_s^2}{t_f(1-\upsilon_s)} \tag{5.3}$$

where $1/R$ is the curvature, E_s is the Young's modulus of the substrate, t_s is the substrate thickness vs is the Poisson's ratio of substrate, and t_f is the deposited film thickness.

Both X-direction and Y-direction stresses are measured and mean is taken to find out the actual stress.

5.3 Annealing Effects

Annealing treatment is given to the wafers in furnace involving treatment of wafers at a high temperature in nitrogen ambient at 600 °C. The ramp rate and cooling rate to be optimized and generally ramp up rate 10 °C/min and cooling rate of 3–5 °C/min is kept for better performance. Two wafers having layer thicknesses from 5000 Å to 22,000 Å are deposited and initial stresses are measured.

(i) Effect of Layer Thickness

Figure 5.5 shows a decrease in the stress as we increase the thickness of the film. This observed behavior is in accordance with literature.

(ii) Effect of Annealing Temperature on Doped Oxide Layer

It is well established that the stress in deposited thin films decreases with annealing as observed experimentally. This is attributed to the fact that while temperature is ramping up, the atoms of the thin film get enough energy to flow, and while slow ramp down rate, the atoms have enough time to settle down into low energy states. Residual stress basically represents the extra energy stored in the thin film due to rapid cooling that happened after the PECVD deposition. The slow ramp down step of annealing provides a way to release all this extra energy and lower the stress.

Although there have been many studies exploring the role of annealing, but stress effect due to annealing temperature variation is presented in this section [5–7]. The Fig. 5.6 demonstrates practically change in stress due to temperature variation.

As observed, with the increase in annealing temperature the stress reduction is higher. This may be attributed to the fact that atoms get higher energy to flow, and hence can settle in low energy states. Analysis of data point for 20,000 Å thickness, it can be inferred that keeping wafer at lower temperature for longer time has almost the same effect as keeping the wafer at higher temperature for less time. The data point for 22,000 Å thickness (Fig. 5.7), it can be established that longer annealing time has an effect of lowering the stress.

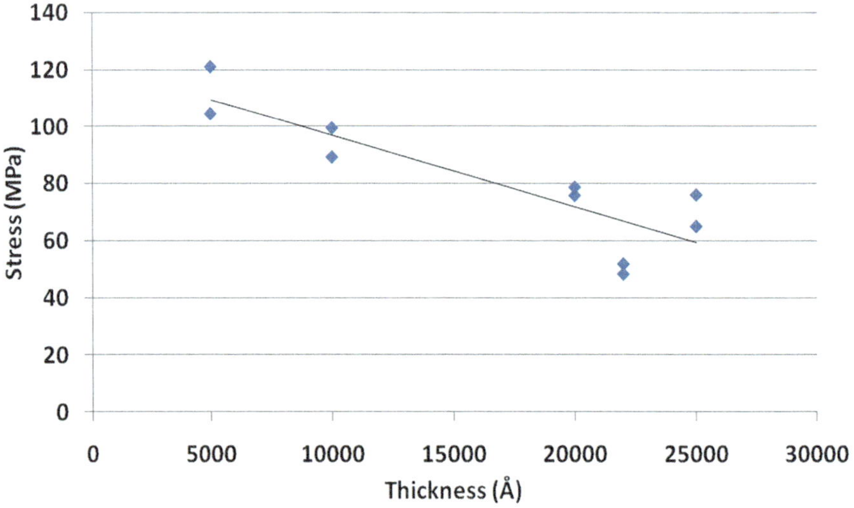

Fig. 5.5 Thickness versus stress graph for PSG oxide

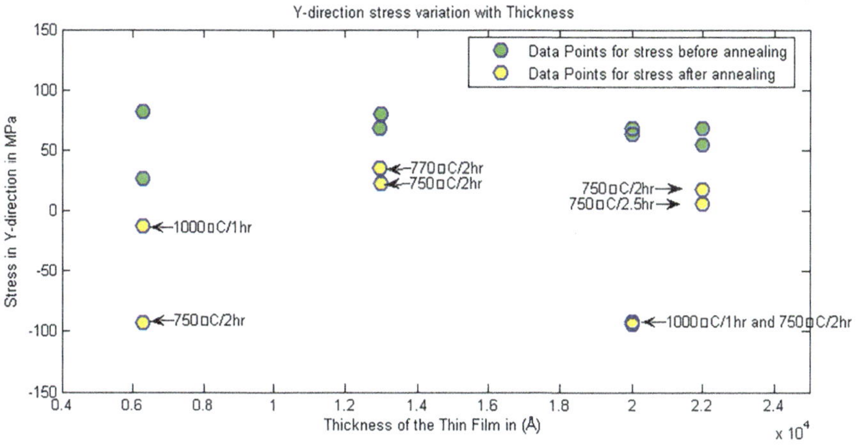

Fig. 5.6 Stress data points before and after annealing

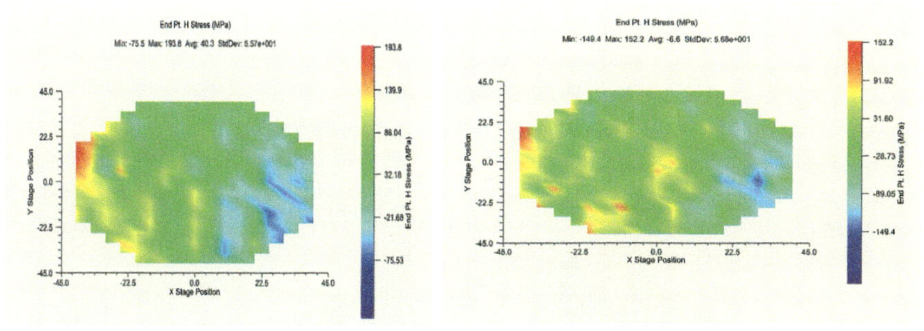

Before Anneal (As deposited) **After Anneal (770°C/2hrs)**

Fig. 5.7 2D stress map before and after annealing

(iii) Effect of Annealing Temperature on PECVD Nitride Layers

Stresses reduce with increase in thickness and as observed from the Fig. 5.8, the magnitude of stress decreases with increase in thickness of the thin film.

Table 5.1 represents the measured stress in 500 and 1000 nm thick SiN layer. Oxide thin films also exhibits the same behavior.

5.4 Stresses in ZnO Layer

Keeping in view of CMOS compatibility, zinc-oxide (ZnO) is termed as one of the very popularly known piezo-electric thin film. It finds wide applications in the development of commercial MEMS-base acoustic sensor, gas sensor, microphone,

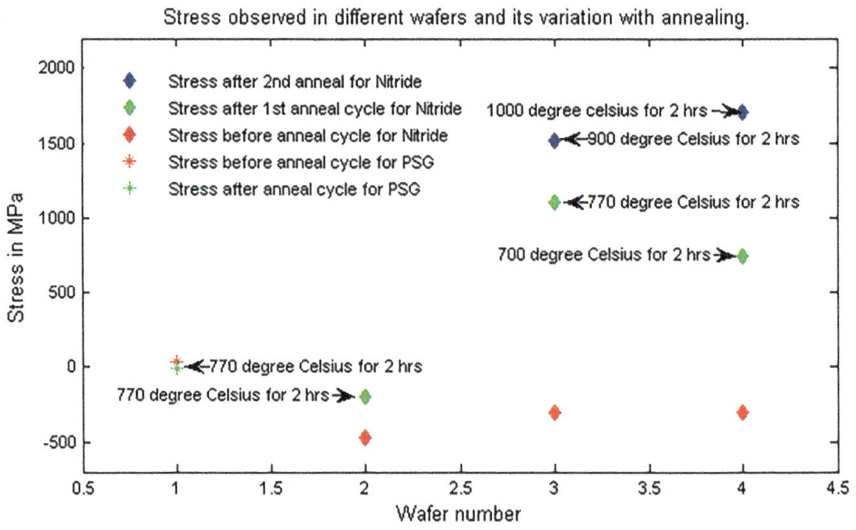

Fig. 5.8 Stress data points before and after annealing for nitride

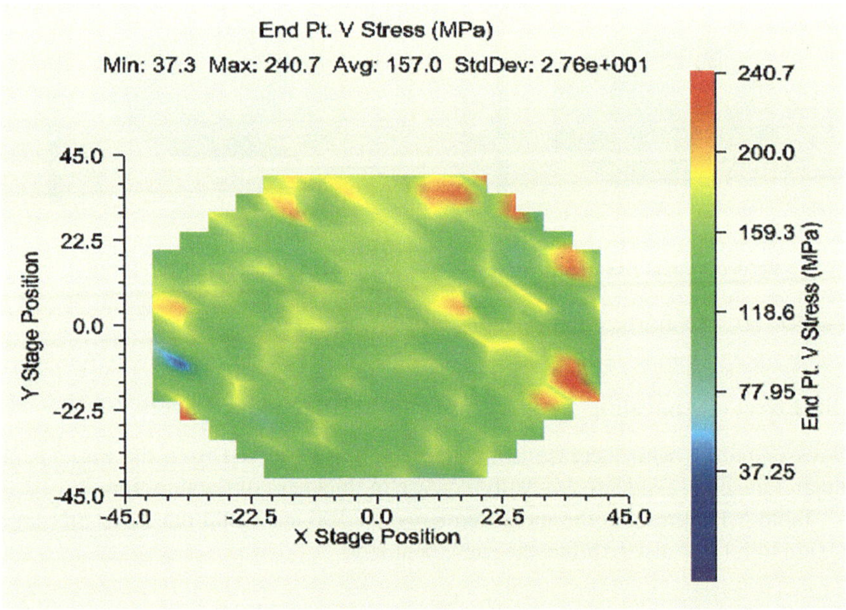

Fig. 5.9 Stress contour

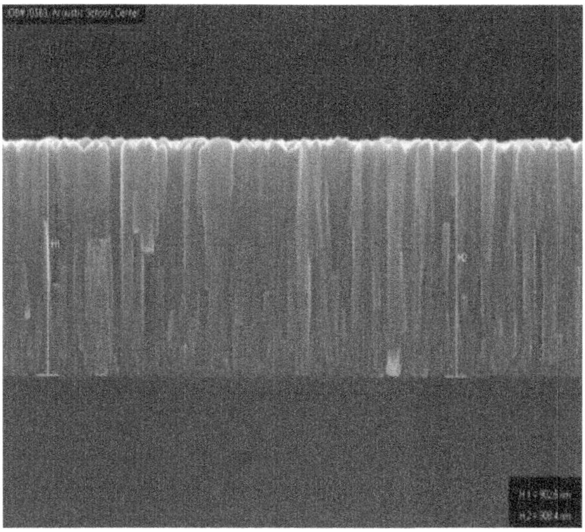

Fig. 5.10 SEM image and surface roughness of the deposited ZnO layer

Table 5.1 Variation of film (SiN) stress with thickness profile

Thickness	Stresses before annealing (MPa)	Stress after annealing (MPa)
500 nm	−475.8	196.3
1000 nm	−747	−300

ultrasonic transducers, etc. These devices are not only extensively used in consumer or commercial on automotive applications, but also gaining popularity in various critical missions of strategic sectors [8–10]. Even, it is also used in Surface Acoustic Wave (SAW) devices. Deposition of ZnO layer depends upon several process parameters, such as temperature, pressure, growth-rate, frequency of the RF-excitation, deposition techniques, and gas chemistry.

Depending upon all these parameters, the quality of the thin film differs. Quality of the film is characterized by the grain size, surface roughness, stability, etc.

ZnO-based sensor development involves various process steps, such as diffusion, wet etching, dry etching, précised lithography, and metallization. Cleaning is obvious in the intermittent states of this fabrication procedure [6]. In this section, we will discuss in detail the effect of various process parameters in the determination of quality of thin films. Further, the effect of humidity and temperature variation of the substrate are studied for the sensor.

5.4.1 Process Parameters

The silicon <100> substrate is chosen here as the base substrate. Using RF-sputtering (magnetron) technique, piezo-electric thin film (ZnO) is deposited. Target size is chosen judiciously for achieving uniform deposition [8, 9]. The prime factors affecting the layer deposition are as follows:

1. Rate of deposition.
2. Temp. of the process.
3. Pressure of the process.
4. Power.
5. Gas chemistry.
6. Target-to-substrate distance.

Various deposition techniques are available. However, sputtering is the preferred category because of its several features, like good adhesion, large area deposition, and requirement of low substrate temperature [10, 11].

Some, pre-sputtering steps are carried out before the actual process. For removing the process contaminations, as a pre-requisite sputtering process is initiated in Argon (An) environment. It ensures the stability of the system. The target-to-substrate distance is generally kept as greater than 10 cum with deposition power of more than 100 W. Argon to oxygen ratio of 60:40 is taken with vacuum level of 10^{-3} to 10^{-5} torr. Deposition-rate is kept as 0.2–0.8 μm/h with a targeted thickness of 1 μm. The purity of the zinc target taken as 99.99% and the pressure is set at 20 mTorr.

5.4.2 *Structural Analysis*

Various characterization techniques are adopted to dig out the quality of the deposited thin film. It can decipher indirectly the molecular structure. Traditionally XRD, SEM, AFM, and FTIR analysis are carried out for the morphological study of the thin films.

Table 5.2 summarizes the techniques and corresponding parameters of various characterization processes.

The fabricated dies are very much proving to moisture of the environment. That's why, with a RH of 95% condition, it is kept within chambers, and FTIR analysis of the die is depicted in Fig. 5.10.

Table 5.2 Characterization techniques and parameters

1	C-axis orientation	XRD
2	Energy band-gap	Optical absorption/STM (STS)
3	FWHM	XRD
4	Grain size	XRD, AFM, SEM
5	Adhesion	SEM
6	Thickness/height	AFM, SEM, Profilometer, reflectance
7	Impurity/contamination	EDAX, XPS, XRD
8	Zn/O ratio (content)	EDAX, XPS

Table 5.3 Film composition

Element	Weight (%)	Atomic (%)
C K_{edge}	2.00	6.24
O K_{edge}	21.15	49.62
Zn L_{edge}	76.86	44.14
Total 100.00		

Further, the crystallographic information of the film can be fetched out with XRD method. As the ZnO film having the hexagonal Wurlzite structure for piezo-electric applications, so the crystal should have C-axis perpendicular to the loose structure [11].

Intensity is determined with 2Q and peak is observed at ~ 34° with low intensity peak of (002). Table 5.3 depicts the composition of the thin film. Full width half maxima pattern (FWHM) is captured in Fig. 5.12. It is due to contamination in the process. FWHM can be improved with the proper cleaning method adopted during the process.

Figure 5.12b clearly indicates the role of contamination in the actual process. Hence, the quality of the film is dictated by process contaminants

Additionally, FTIR response was captured for finding out the transmission characteristics (%) of the film along with dispersion behavior (variation of permittivity with frequency).

Substrate temperature also plays important role in determining film quality. Mathematical equations governing the various parameters of thin films are as follows:

Fig. 5.12XRD plot of ZnO layer a standard b after trials

$$\begin{aligned} t &\sim \frac{0.9\lambda}{\beta\cos\theta} \quad \text{(a)} \\ r &\sim 0.625.\frac{\lambda}{\sin\theta} \quad \text{(b)} \\ d &\sim 0.5\frac{\lambda}{\sin\theta} \quad \text{(c)} \\ g &\sim \frac{\beta}{\tan\theta} \quad \text{(d)} \end{aligned} \tag{5.4}$$

where λ is the wavelength, r is the inter-chain distance, and t is the grain size. Table 5.4 shows the calculated parameters of the ZnO film. Accordingly to the governing equations, the calculated parameters of the thin films are summarized in Table 5.4.

The calculated grain size (~30 nm) is validated practically in the AFM plot (Fig. 5.14). The grain size affects the reflectance and it decreases substantially with the incensement of grain size—with the help of opti-probe, the same process (measuring grain size) can be done. The AFM plot indicates densely packed structure with uniformly distributed grain structures.

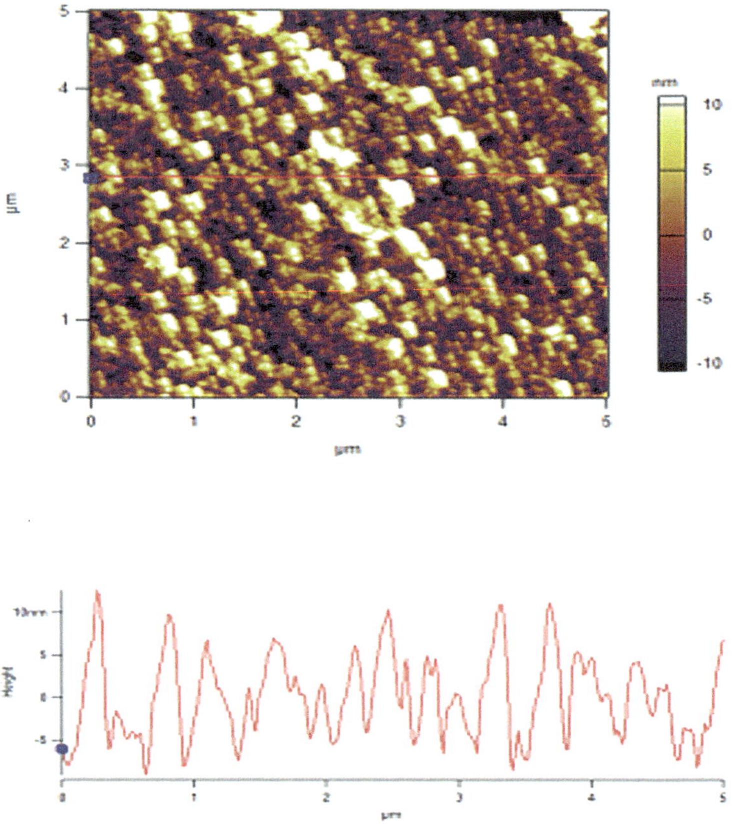

Fig. 5.11 EDAX analysis for the composition of the film (Table 5.3) shows the composition of the film

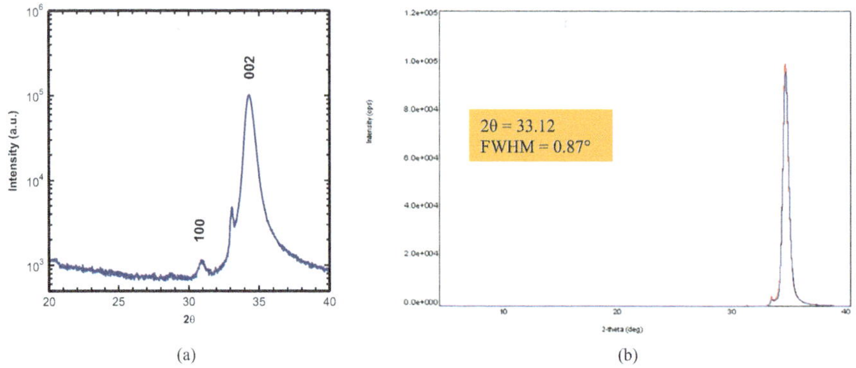

Fig. 5.12 XRD plot of ZnO layer **a** standard **b** after trials

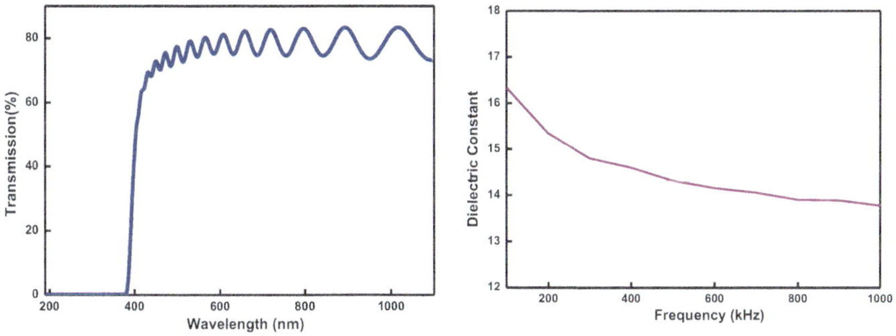

Fig. 5.13 FTIR plot and dielectric constant measurement

Fig. 5.14 AFM plot of the deposited layer

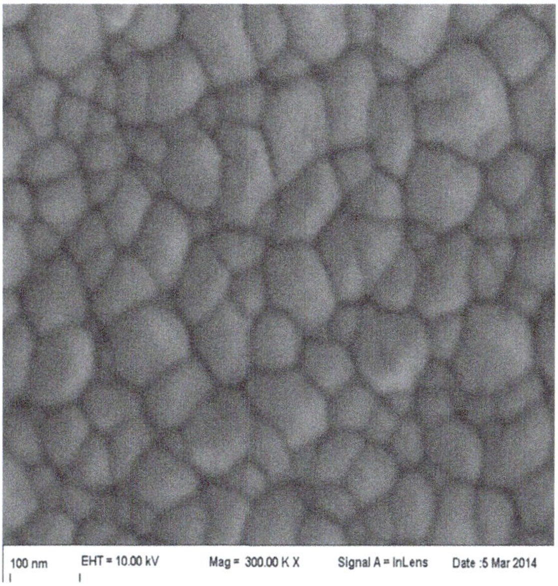

Table 5.4 Calculated parameters for the deposited thin film

S. No.	Parameters	Calculated
1	Grain size (***t***)	30 nm
2	Inter-chain distance (***r***)	1.7 Å
3	Inter-planar distance (***d***)	2.61 Å
4	Distortion parameter (Lattice strain) (**g**)	0.86%

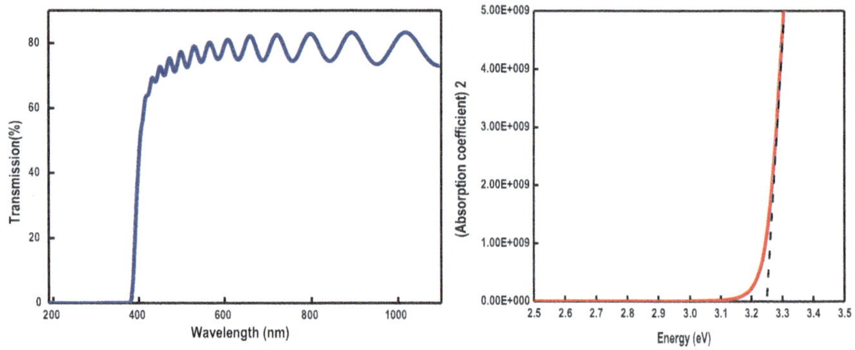

Fig. 5.15 Transmission and absorption coefficient at room temperature

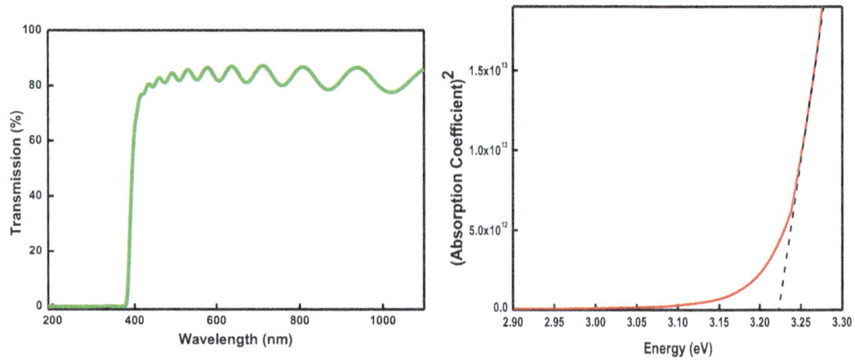

Fig. 5.16 Transmission and absorption coefficient at 300 °C

5.5 Effect of Substrate Temperature

Substrate temperature plays a pivotal role in the quality of deposited thin films. This fact has been experimentally verified with two different conditions, as shown in Figs. 5.15 and 5.16. The stress value of the thin film depends upon the annealing duration and temperature settings [12–15].

5.6 Humidity Response

Effect of relative humidity on the quality of the thin film is studied here. Signature of the moisture effect is captured by FTIR study. Corresponding peaks are observed depending upon the electro-negatively of back-bond. Figure 5.17 depicts the FTIR plots, deciphering the moisture effect on the thin film. Due to dipole–dipole interactions, there is a sudden change in the spectrum is noticed [16–21].

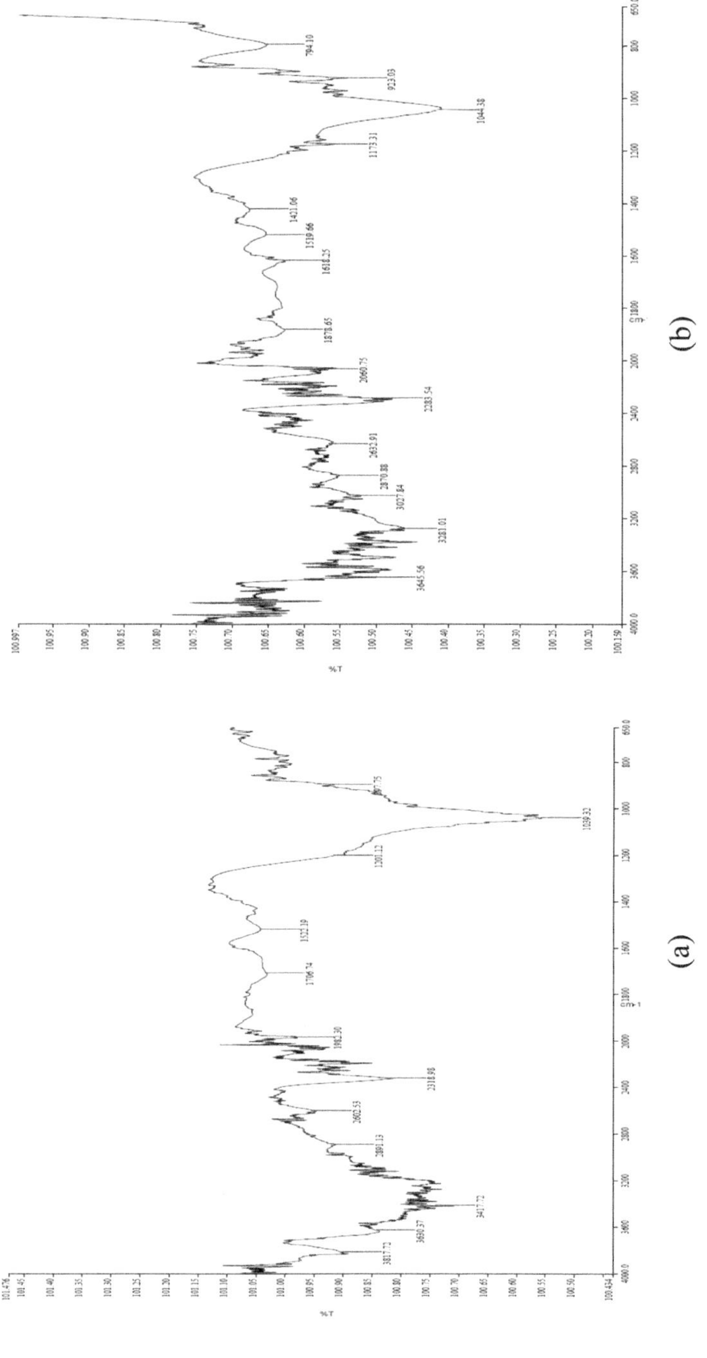

Fig. 5.17 Moisture effect on layer **a** without and **b** with humidity

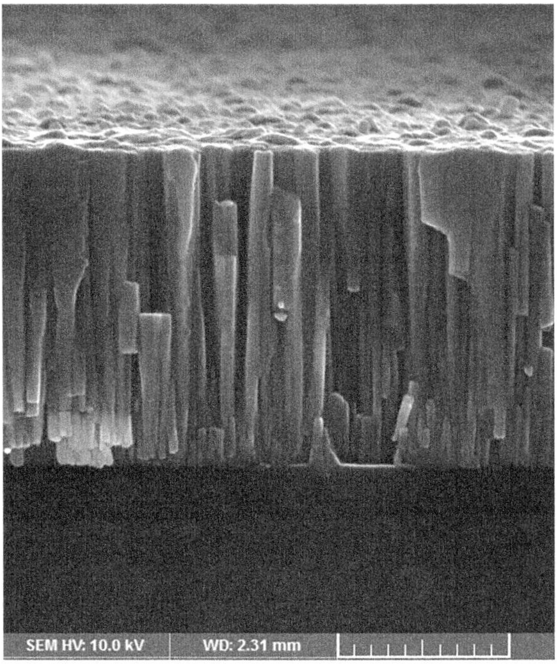

Fig. 5.18 SEM plot after etching

With the etch-rate evaluation, the quality of the thin film can be predicted in other way. Standard wet etch chemistry for ZnO film is a combination of NH_4Cl: CH_3COOH: H_2O and the etch-rate is ~ 2.95 μm/min. Any deviation (small or large) can indicate the degradation of the quality of thin film.

Another figure of merit of thin film is the evaluation of its dissipation factor (DF) or loss tangent. Its standard value is ~ 0.001, which denotes a good quality of thin film usable for acoustic sensor or any capacitive sensor-based application with a very low dielectric loss. Moreover, the annealing process at the end of film deposition leads to a good quality of thin film as desired.

5.7 Summarization

This chapter outlines the deposition of piezo-electric thin film (ZnO) and its various characterization procedures. Factors affecting the deposition process of the films are discussed in detail. Prime reasons of inducing stress in the thin film have also been investigated. Analytically as well as experimentally, causes were found which determine the quality of ZnO film. Various experimental data were presented to inspect the effect of substrate temperature, effect of annealing, humidity/moisture effect, etc. Even the contamination incurred in the process can alter the yield has also been highlighted here in this chapter.

Dispersion characteristics as well as film transmission pattern with the variation of operating frequency or wavelength have also been studied in detail. To address the reliability aspect, the inadvertent residual stress within the thin film is also studied.

Practical aspects of using thin film-based MEMS sensors in many multidisciplinary applications such as acoustic sensor, μ-phone, gas sensor, and ultrasonic transducer; FBAR, SAW devices, have been demonstrated here.

Question

1. What is thin film?
2. Why the stress is developed in thin film?
3. Define stress, strain with respect to thin film.
4. What are the ways to get rid of stress within the film?
5. How to measure the stress in thin film?
6. How the annealing affect the residual stress with the thin film?
7. How the stress profile in thin film may change with the process parameters?

References

1. Mattrox DM (2001) Atomistic film growth and resulting film properties: residual film stress. Vacuum Technol Coat 22–23
2. Guckel H, Randazzo T, Burns DW (1985) A simple technique for the determination of mechanical strain in thin films with applications to polysilicon. J Appl Phys 57(5):1671–1675
3. van Drieenhuizen BP et al (1993) Comparison of techniques for measuring both compressive and tensile stress in thin films. Sens Actuators A37–38:756–765
4. Tarraf A, Daleiden J et al (2004) Stress investigation of PECVD dielectric layers for advanced optical MEMS. J Micromech Microeng 14:317–323
5. Huang H et al Evaluation of elastic modulus and stress gradient of PECVD silicon nitride thin films. Retrieved from http://eprint.uq.edu.au/archive/00000836
6. Huang H (2006) Effect of deposition conditions on mechanical properties of low-temperature PECVD silicon nitride films. Mater Sci Eng A435–36:453–459
7. Zhang X, Chen K-S, Spearing SM (2002) Residual Stress and fracture of thick dielectric films for power MEMS. Appl IEEE 164–167
8. Elbrecht L et al (1997) Comparison of stress measurement techniques in surface micromachining. J Micromech Microeng 7:151–154
9. Prasad M, Shukla V, Khanna VK (2013) Design and fabrication of Si diaphgram ZnO piezo electric film based MEMS acoustic sensor using SOI wafers. IEEE Trans Semiconductor Manuf **26**(2):233–236
10. Singh K (2015) Development of indigenous metal oxide gas sensors with embedded heater using CMOS process. Punjab Acad Sci J **13–14**(1 & 2):1–4
11. Ananthsuresh GK et al (2005) Micro and smart systems. Wiley India
12. Huang W (2012) C-axis ZnO piezoelectric thin film prepared by RF magnetron sputtering for SAW filters. In: 2nd international conference on computer application and system modelling, pp 217–220
13. Zaheer, N, Singh K (2016) The challenge of convex corner undercutting phenomena observed in silicon micromachining and its circumvention. Technical Hindi Seminar, SCL, Mohali
14. Singh K (2017) Contaminations in MEMS processes and removal methodology. ICTACT J Microelectron 3(1):375–378

15. Ozgur U, Alevov YI, Lui C, Teke A, Reshchikov MA, Dogan S, Avrutin V, Cho S-J, Morkeel H (2005) A comprehensive review of ZnO materials and devices. J Appl Phys 98
16. Arora A, Arora A, Dwivedi VK, George PJ, Srrnivas K, Gupta V (2008) Zinc oxide thin film based MEMS acoustic sensor with tunnel compensation. Sens Actuators, A141:256–261
17. Ondo Nelong R, Omanda HM et al (2013) Piezoelectric ZnO thin films by RF magnetron. Int J Eng Sci (IJES), 2(12):58–64
18. Cheng D-L, Kao K-S, Liang C-H, Wang Y-C et al (2017) Piezo electric response evaluation of ZnO thin film prepared by RF magnetron sputtering. 2nd ICMSNT, MATEC web of conferences, vol 109
19. Xia X-G, Yuan Y, Du H, Liu B (2013) Study of piezoelectric ZnO thin films for contact sensing and head actuation. IEEE Trans Magnet 49(6):2539–2543
20. Shinde SD, Patil GE, Kayale DD, Wagh VG, Gailkwad VB, Jain GH (2012) Effect of annealing on gas sensing performance of nano-structured ZnO thick film resistors. Int J Smart Sens Intell Syst 5(1):277–290
21. Singh K, Nirmal AV (2018) Annealing effects on stress behavior for PECVD deposited Oxide and Nitride thin layers for MEMS applications. AeCND, Bengaluru, India

Chapter 6
Micro-devices: Planar Heater

This Chapter outlines the development of a microheater structure on silicon as well as glass substrates. The structure is a planar type and miniaturized form. Standard CMOS-fabrication steps have been implemented to realize it. This Chapter details about the practical considerations, bottlenecks, and ways to mitigate it. Electrical characterization is also briefed herewith-detailed data analysis. Further, novel-packaging techniques are summarized here keeping in view of all practical constraints.

6.1 Introduction

Now-a-days microheater structure is preferably developed using MEMS technology, which offers its several merits viz. compact shape and size, very low power consumption, less thermal mars, low unit cost, capability of batch processing. In recent times, the usage of microheater is gigantically increasing in several domains, like relative humidity (RH) sensor, flow sensor, metal-oxide-based gas sensors, etc. Because of its low power consumption feature, its popularity is increasing day-by-day in multifaceted sensor applications. Prime features of microheater designs are as follows:

1. Low thermal loss.
2. Uniform distribution pattern of temperature over membrane.
3. Significant thermal isolation of the heater structure from the surroundings.
4. Very low thermal response time.
5. Very low power consumption.
6. Provision of scalability feature.

The designer has to optimize all these prime parameter while designing microheater as per the required specifications.

K. Singh and A. Karmakar, *Micro Electro Mechanical System (MEMS) Technology for Multifaceted Applications*, https://doi.org/10.1007/978-981-96-6162-6_6

Conventionally, the shapes of microheaters are: meander type, S-shape, spiral, honeycomb, fan-shaped, elliptical, drive-wheel, etc. [1]. Shape of the heater greatly relies on the end usage. However, the selection of proper material depends on several factors, such as thermal conductivity, electrical conductivity or resistivity, heat capacity, compatibility of the material with standard wafer (or substrate) of the micro-fabrication technology, etc. For example, metal-oxide-based 'Gas sensor' topology generally requires temperature rise of around 300–400 °C [2] within the specified time frame whereas RH sensor need to generate a temperature less than 100 °C [3].

6.2 Design and Development of Microheater

As an input, electrical current is applied to the microheater structure and as a result, it produces heat energy. The produced heat is directly proportional to the square of current density applied to it. Mathematically, it can be expressed as follows [4]:

$$H \,\alpha\, J^2 \tag{6.1}$$

where H = generated heat through microheater structure (J).

J = Electrical current density (A/m^2).

Now,

$$J = \sigma E^2 \tag{6.1a}$$

where σ = electrical conductivity of the heating element (S/m).

E = Magnitude of the applied electric field (V/m).

Resistance of the microheater structure depends on the length, breadth, thickness of the structure, along with its material compositions and thermal mass.

Semiconductor based gas sensor requires metal-oxide layer for absorption to detect the presence of various gases. Material selection is an important part in microheater fabrication. Mostly microheaters are fabricated using Poly silicon, Titanium, and Platinum. Material properties are given here in Table 6.1.

Table 6.1 Comparative study of material properties for various microheater elements

No	Properties	Titanium	Platinum	Poly silicon
1	Thermal conductivity (W/cm/C)	0.219	0.716	0.148
2	Electrical resistivity (Ω.cm)	4.8e−5	1.06e−5	82
3	Thermal expansion coeff. (K^{-1})	$8.6 * 10^{-6}$	$8.8 * 10^{-6}$	$4.7 * 10^{-6}$
4	Specific heat (J/g/C)	0.523	0.133	0.100
5	Density (g/cm^3)	4.51	21.45	2.32

Microheater is based on the application requirements and gas sensor good temperature uniformity all over the membrane and the substrate temperature to be achieved is around 250–350 °C. On another side, pyro igniter microheater should be able to attain ~ 450 °C within 100 ms. For that, different types of microheaters are to be designed and characterized which are presented in this section.

6.2.1 Power Compensated Design Microheater

This microheater was fabricated on silicon wafer and specially designed for gas sensor. A microheater should have very good heat uniformity for good gas sensing property of metal-oxide sensing layer. Design of this microheater is optimized to attain 250–350 °C temperature at low voltage, fast thermal response time, low power consumption, low thermal mass, and temperature uniformity over membrane.

In conventional microheater, the temperature of membrane tends to decrease dramatically away from the center of the membrane. The reason for this tendency is the greater heat loss resulting at the outer areas of the heater. To achieve this increase in heat generation it is necessary to increase the power consumption at the edge of membrane where the majority heat is lost. Power compensated design in planar type microheaters enhances temperature uniformity and increases the heating area, which is particularly useful for applications like semiconductor gas sensors. This approach aims to distribute power more evenly across the heating element, mitigating temperature variations that can occur with standard designs. Therefore, in the X-direction, the outer area has a smaller and denser pattern than center in order to increase the resistance and hence temperature.

It was packaged on TO-8 as this type of packaging can sustain very high temperature. A 675 μm thick Si-wafer was used as substrate. Other layers were 0.05 μm Oxide, 0.3 μm Nitride, and 0.6 μm Titanium. Attained temperature or microheater was measured by mounting an IST sensor as shown in above Fig. 6.1.

This microheater got breakdown after attaining 416 °C at 13.5 V. This microheater takes ~ 90 s. to attain maximum temperature. Temperature stability test was carried out and above microheater was stable for around 6 h at a fixed voltage with fixed temperature. Simulated results show good heat uniformity over the membrane. Measured temperature values are given here in table. Measured results are presented here in Table 6.2 as well as in Figs. 6.2 and 6.3 (Table 6.3).

Features

I. Current and power consumption is very high in above design. There are 12 resistances in parallel combination so the equivalent resistance is very less (~10Ω). A new design with 8 resistances in parallel is in process to increase microheater resistance.
II. Heater surface area is much larger that causing heat dissipation through the bottom silicon substrate. Modified heater results are given in next sections.

Fig. 6.1 Heater structure (power compensated design)

Table 6.2 Test results data for power compensated design

Supplied voltage (V)	Temperature attained (°C)	Power consumption (Watts)	Resistance (Ω)
2	51	0.22	18
3	78	0.43	20
4	112	0.72	22
5	141	1	25
6	174	1.3	27
7	211	1.5	28
8	240	2.1	30
9	283	2.5	31
10	314	3	33

III. It was observed during simulation that microheater attained temperature got increases around 6–7 times by bulk micromachining of Si substrate. Mask fabrication and design are under process to perform back side bulk micro machining so that high temperature can be obtained at low voltage and low power consumption.

IV. Platinum has good response time than Titanium. Simulation and literature survey supports that platinum is good metal for microheater. Masks are under process for this trial as Pt need reverse polarity mask for liftoff purpose.

V. Microheater temperature uniformity test to be carryout using IR camera at SAC or IISc so that simulated results can be verified (Fig. 6.4).

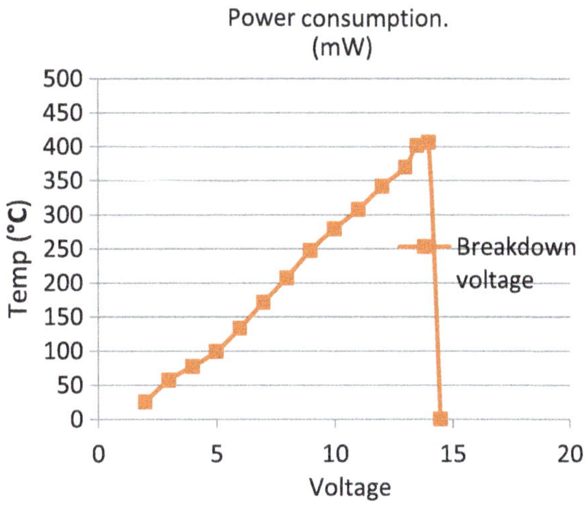

Fig. 6.2 Attained temp. versus applied voltage plot

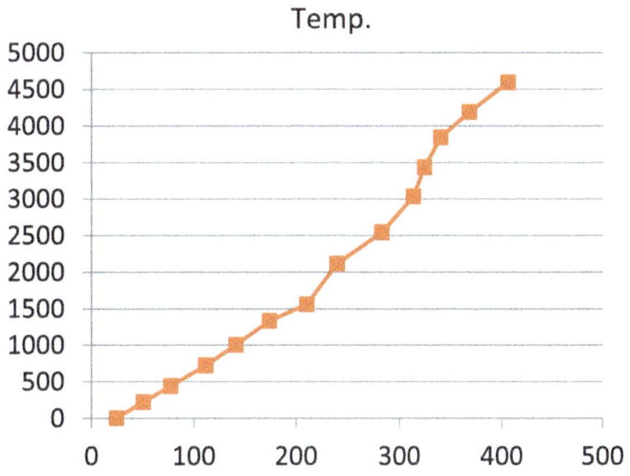

Fig. 6.3 Power consumption versus attained temp. plot

6.2.2 Spiral Shape Microheater

This microheater was fabricated on silicon wafer and it was designed for gas sensor. Power compensated technique was applied in this microheater also as temperature uniformity is the main concern in gas sensor. For that the outer structure of micro-heater has less width in order to increase the resistance and hence temperature. Layers thickness is kept same for spiral microheater as in the case of power compensated design structure and both microheaters were fabricated on same wafer using same mask plate. It is proposed to fabricate Titanium microheater and resistance was ~ 120Ω. Simulation results show good heat uniformity over membrane as shown in below Fig. 6.5.

Die size for above microheater is kept $1.2 \times 1.2\ \text{mm}^2$, and the actual heater area is around $562 \times 576\ \mu\text{m}^2$. It was finalized based on literature survey that microheater

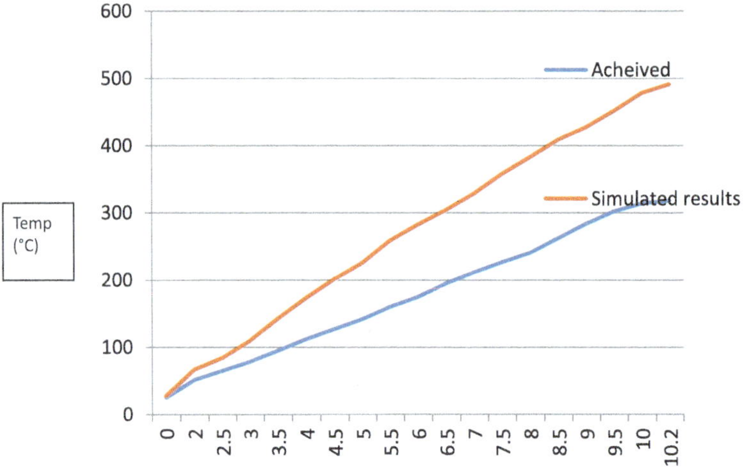

Fig. 6.4 Simulated and achieved result for power compensated design microheater

Fig. 6.5 Spiral microheater design and dimension

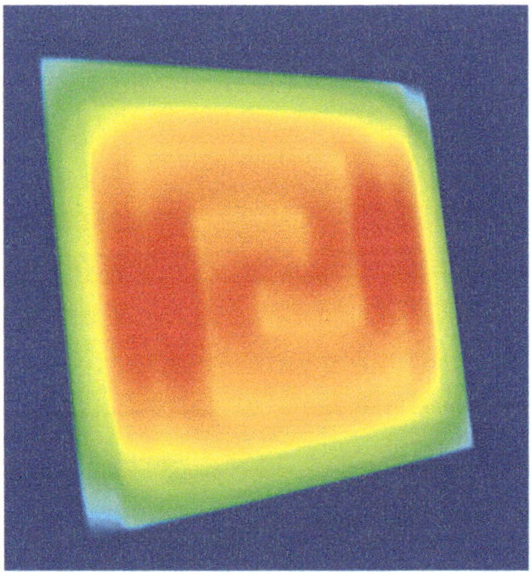

to die size should have 1:2 ratio. This microheater achieved maximum 149 °C at 21 V. The advantage of this microheater is that it has very less power consumption (300–900 mW) and current consumption (~60 mA). This microheater takes ~ 50 s. to attain maximum temperature. Temperature stability test was carried out and above microheater was stable for around 5 h at a fixed voltage with fixed temperature. Simulated results show good heat uniformity over the membrane (Figs. 6.6 and 6.7).

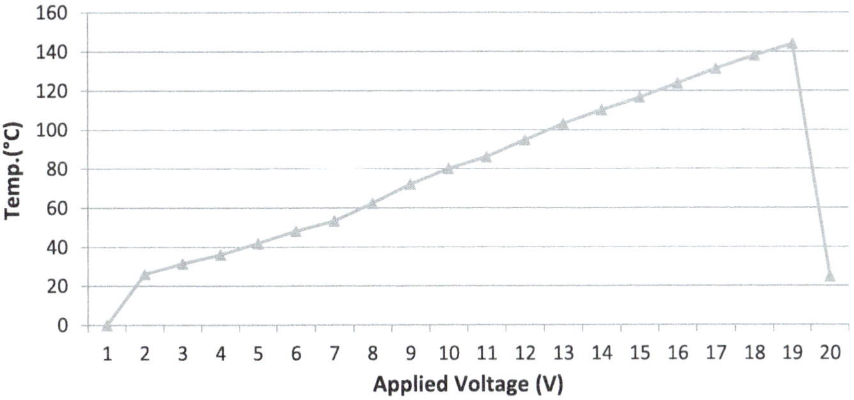

Fig. 6.6 Achieved temperature. versus applied voltage (*Courtsey* SCL)

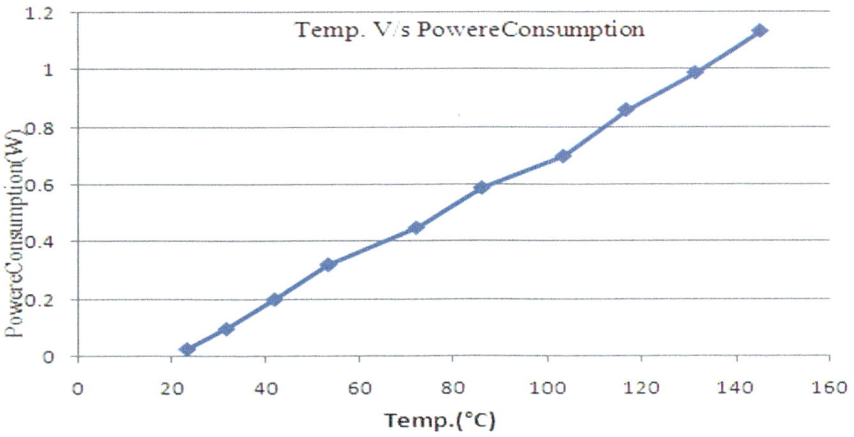

Fig. 6.7 Power consumption versus achieved temperature plot (*Courtsey* SCL)

6.2.3 Platinum Microheater

This microheater was fabricated on glass wafer and specially designed for pyro igniter with 100 Ω resistance. Pyro igniter microheater should have capability to attain 450–500 °C temperature within 100 mS 500 μm thick glass substrate was used to fulfill that requirement as glass has very less thermal conductivity. Actual mask was prepared for 0.8 μm Platinum but later we were able to deposit only 0.2 μm Platinum. The resultant resistance of microheater was ~ 750 Ω instead of 100 Ω. This was a single layer process. Testing of this microheater was carried out at VSSC, Trivandrum and SCL, Chandigarh. This type of microheater was able to ignite ZPP (igniting material) within 100 ms (Figs. 6.8 and 6.19).

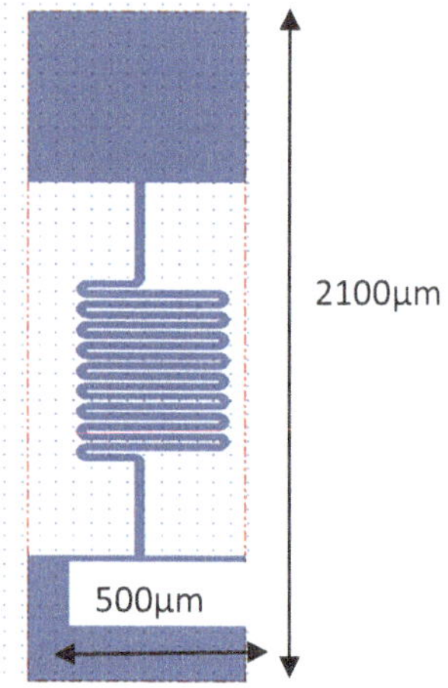

Fig. 6.8 Platinum spiral structure

Fig. 6.9 Packaged die for testing

This microheater was able to attain 110 °C at 36 V. Microheater wire bonding breaks down beyond 36 V as it was COB bonding. Few trials were carried out on TO-8 package (Fig. 6.10).

Few other type of testing was carried out using this microheater as it will be used in Pyro igniter. It was observed at VSSC that practically measured current does not match with theoretically estimated current after 40 V. VSSC test results are given here for few dies (Tables 6.4, 6.5, 6.6 and 6.7).

Above testing results were cross verified with the testing results are given here for 2 dies which were packaged on TO-8.

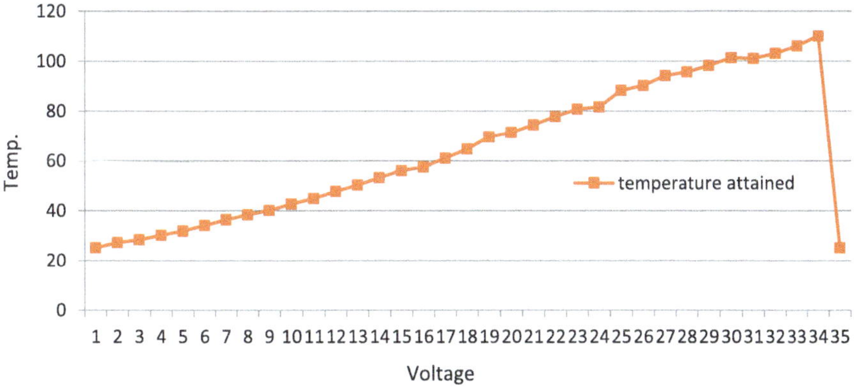

Fig. 6.10 Platinum microheater substrate temp. versus applied voltage graph

Table 6.3 Test results of spiral microheater

Supplied voltage (V)	Temperature attained (°C)	Power consumption (Watts)	Resistance (Ω)
2	23.10	0.026	150
4	31.56	0.098	163
6	41.98	0.201	179
8	53.27	0.321	199
10	72.03	0.450	222
12	86.01	0.590	244
14	103.22	0.700	280
16	116.54	0.861	297
18	131.33	0.990	327
20	144.98	1.136	352

Table 6.4 Test results of pyro igniter

S. No.	Resistance of die (Ω)	Applied voltage (V)	Expected current (mA)	Observed current (mA)
1.	764	20	26	29.5
2.	753	30	39.8	58.5
3.	745	35	46.9	75
4.	740	40	54	102

Die 1: Resistance (@ambient temp.) = 666Ω.

Die 2: Resistance (@ambient temp.) = 810 Ω.

In reverse process, a die(R = 570 Ω) was placed in a chamber and temperature was increased. The measured resistance values are given here:

Table 6.5 Current and resistance variation with applied voltage

Applied voltage (V)	Measured current (mA)		Theoretical current (mA)	Measured resistance(Ω)
	Initial	Final		
15	15.15	14.56	21.2	705
20	17.31	16.79	26.7	747
25	19.0	18.52	32.2	775
30	20.4	19.87	36.8	815
32	20.82	20.4	38.7	825

Table 6.6 Current and resistance variation with applied voltage

Applied voltage (V)	Measured current (mA)		Theoretical current (mA)	Measured resistance(Ω)
	Initial	Final		
15	13.0	12.65	17.3	865
20	15.0	14.65	22.2	900
25	16.6	16.25	27.2	925
30	18.0	17.5	31.1	960
32	18.4	18.0	32.9	970

Table 6.7 Microheater resistance variation with chamber temperature

Chamber temp. (°C)	Heater resistance (Ω)
25	561
50	597
75	637
100	672
125	709

Conclusion of testing:-

- In each case initial current is higher than the initial current of previous voltage. After stabilizing (after 6–7 min.) current goes down as the resistance got increase due to self-heating of microheater.
- In each case the resistance of heater gets increase if we are increasing applied voltage.
- In earlier testing of microheater the obtained current was 29.5 mA @ 20 V and 58.5 mA @ 30 V which is much higher than our testing results.
- Previous testing results show that this heater can achieve around 130 °C temperature at 32 V. The net increase in the resistance of microheater is 170–200 Ω by applying 32 V/125 °C. This can be verified by above two different approaches.

Design of above microheater was modified for 100 Ω resistance using 0.2 μm thick Platinum. Achieved resistance of fabricated microheater is ~ 106 Ω and it was able to achieve 207 °C@ 21 V. Current behavior for initial 180mS of this microheater was also captured using oscilloscope (Fig. 6.11, 6.12, 6.13, 6.14, 6.15 and 6.16).

Fig. 6.11 Microheater design

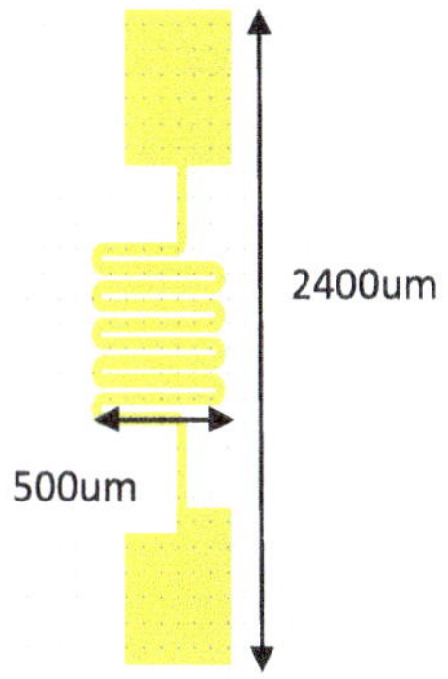

Fig. 6.12 Temp. versus applied voltage

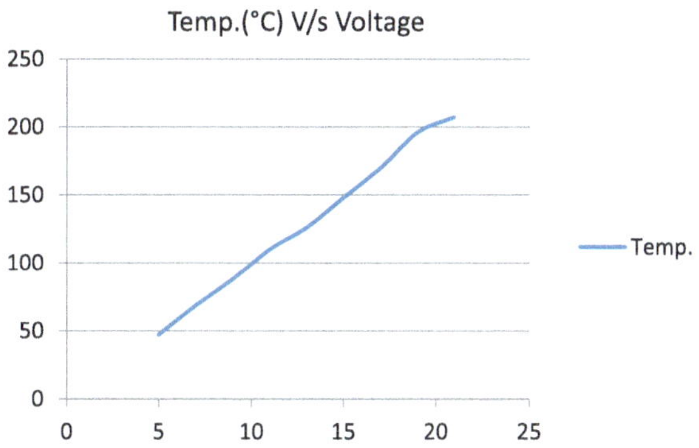

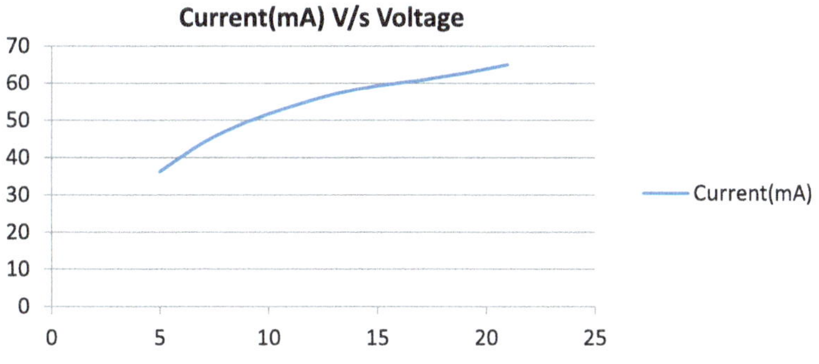

Fig. 6.13 Current variation with applied voltage

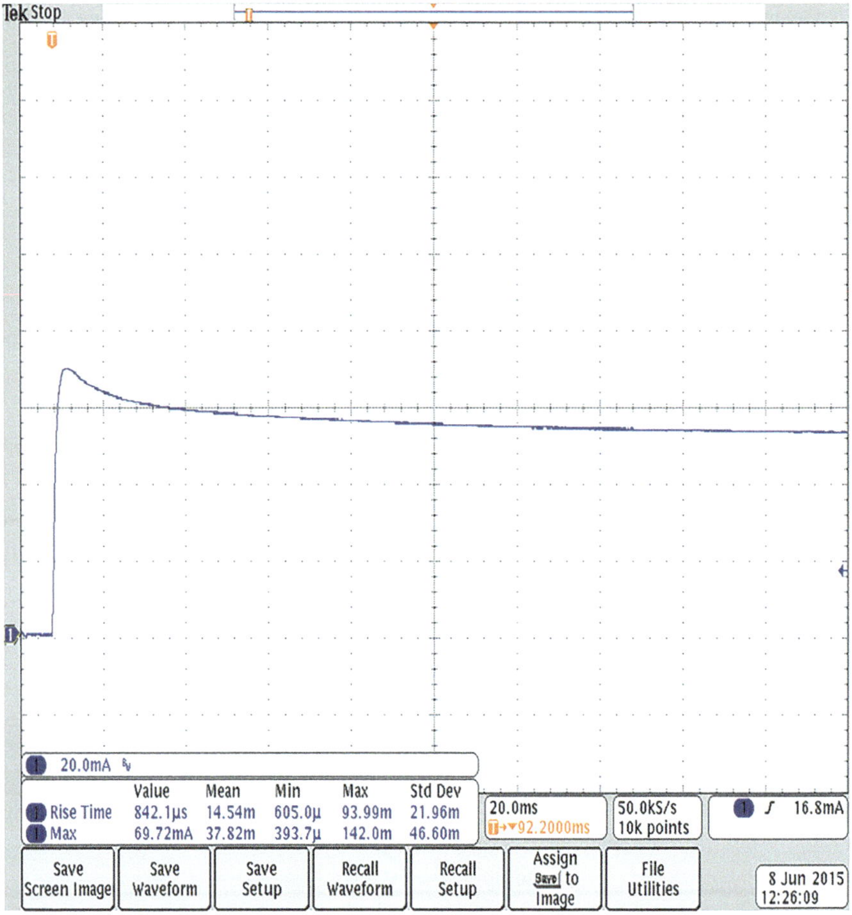

Fig. 6.14 Current behavior at 10 V

Above plot shows this microheater takes 40–50 ms for getting stabilize known as settling time. Initial peak was observed in all trials. Settling time is the time taken to stabilize the current for any given electronic device.

6.2.4 Pt/Al Microheater

This microheater was fabricated for temperature sensor on silicon wafer with Oxide/Nitride stacks. Platinum thickness was 1180 Å and pads were fabricated using 0.6 µm Al thickness. Theses microheater have ~ 1200Ω resistance. These micro-heaters were packaged on TO-8 package and testing was carried out to measure substrate temperature (Fig. 6.17).

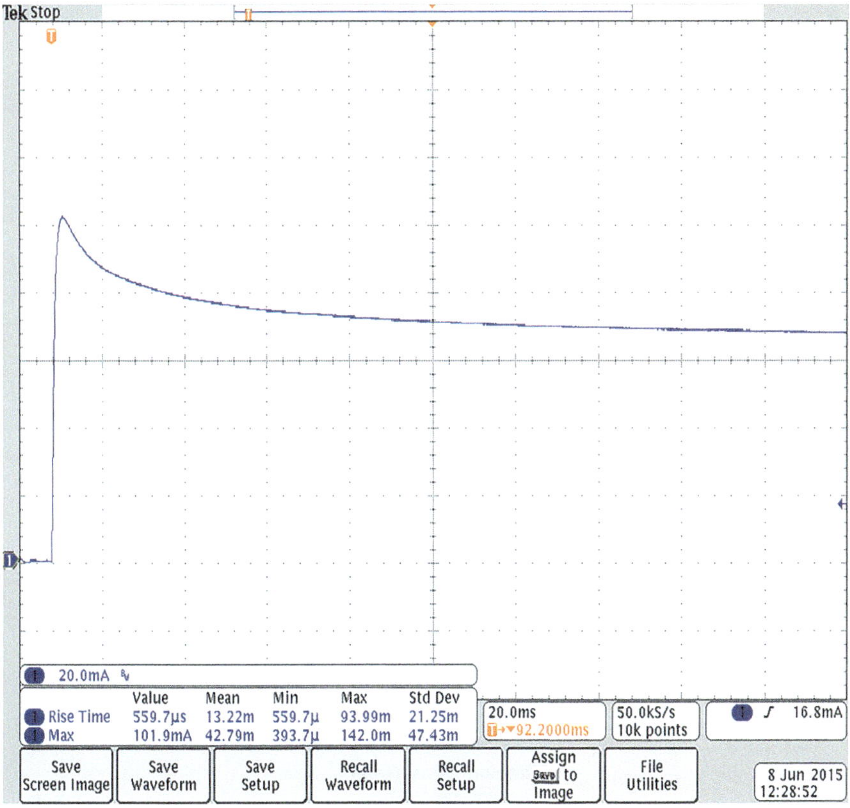

Fig. 6.15 Current behavior at 20 V

This microheater was able to attain maximum 220 °C substrate temperature at 55 V. It was observed that microheater got breakdown beyond 55 V. The maximum current consumption was ~ 35 mA. Bulk micromachining of above microheater can give better results (Table 6.8).

6.2.5 Microheater Fabrication on Glass Wafer

Six different types of microheater are designed for pyro igniter and gas sensor. These microheaters were fabricated on glass wafer (0.7 mm thick). Platinum (0.22 µm thick) was used as heater material. As glass has very less thermal conductivity so this type microheater achieves relatively high temperature than silicon substrate. Testing results are given here for different designs.

1. **Modified Power Compensated Design**:- Earlier power compensated design had 12 resistive elements in parallel to each other. The resultant resistance was very

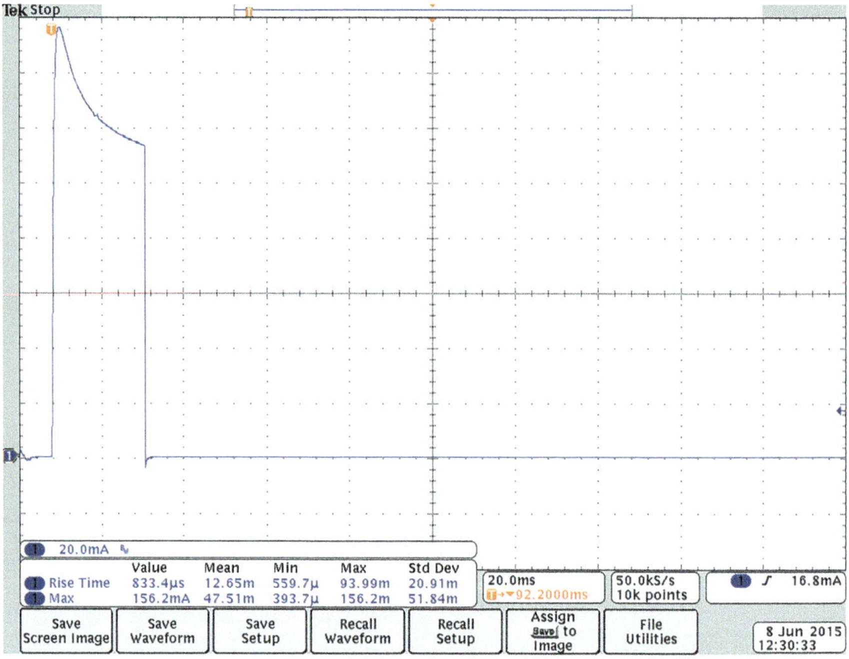

Fig. 6.16 Current behavior at 50 V (*Courtsey* SCL)

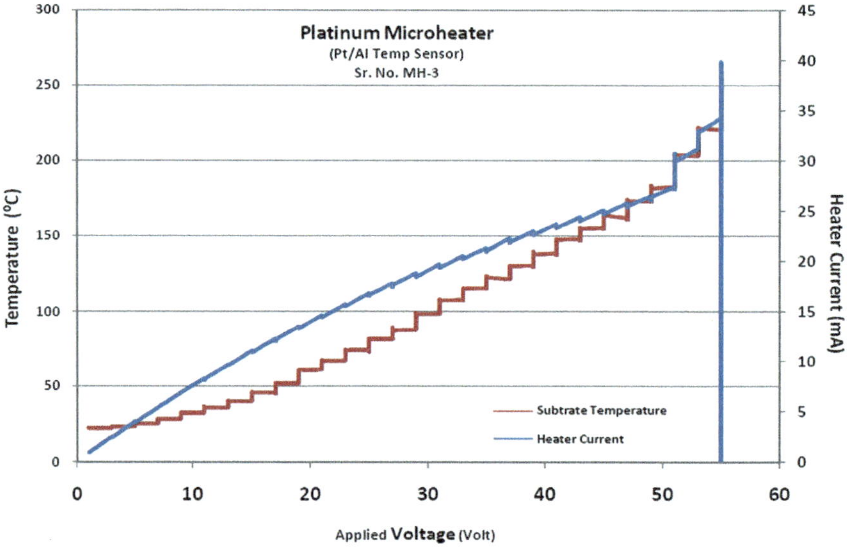

Fig. 6.17 Substrate temperature and current variation with applied voltage (*Courtsey* SCL)

Table 6.8 Summary of all above microheaters

No	Design	Resistance (Ohm)	Dimension (mm)	Breakdown voltage (V)	Attained Temp. (°C)	Max. power consumption (W)
1	Power Comp. design	19	1.2 * 1.2	15	400	3.0
2	Spiral	120	0.6 * 0.6	20	149	1.1
3	Platinum	750	2.1 * 0.5	36	110	1.72
4	Platinum	106	2.4 * 0.5	21	207	1.3
5	Pt/Al	1200	1.2 * 1.1	55	220	2.52

less (~19 Ω) causing very high power consumption. This design was modified and now it has 8 resistive elements in parallel and it was fabricated on glass wafer. The resultant resistance is ~ 24 Ω for the design given here in below Fig. 6.18.

This microheater was able to achieve 172°V@7V. It gets breakdown beyond 7V. Current variation with applied voltage is given below.

2. **Hotplate Microheater**:- That hotplate was designed on glass wafer for gas sensor having 36Ω resistance. The maximum achieved substrate temperature was 258 °C@12 V and gets breakdown beyond 12 V.

Current variation with applied voltage for hot plate is plotted here:

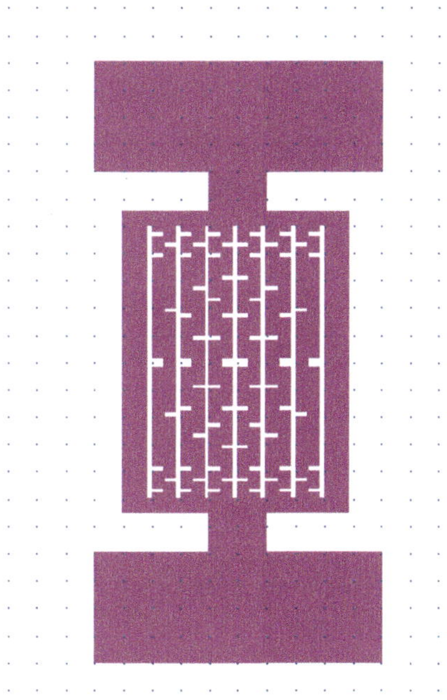

Fig. 6.18 Modified power compensated design

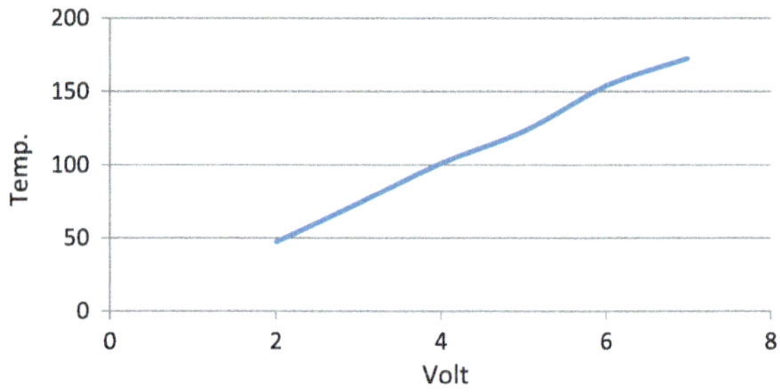

Fig. 6.19 Temp. variation with applied voltage

3. **Serpentine Shape Microheater**:- This microheater was able to achieve high temperature during testing and can be useful in pyro igniter. The resistance of this microheater was 145 Ω and it was able to achieve 283 °C@34 V. Plots are given here for different testing (Figs. 6.19, 6.20, 6.21, 6.22, 6.23, 6.24, 6.25, 6.26, 6.27, 6.28 and 6.29).

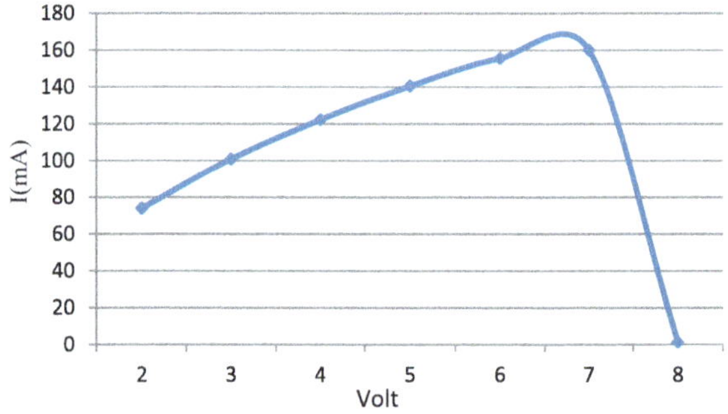

Fig. 6.20 Current variation with applied voltage for power compensated design

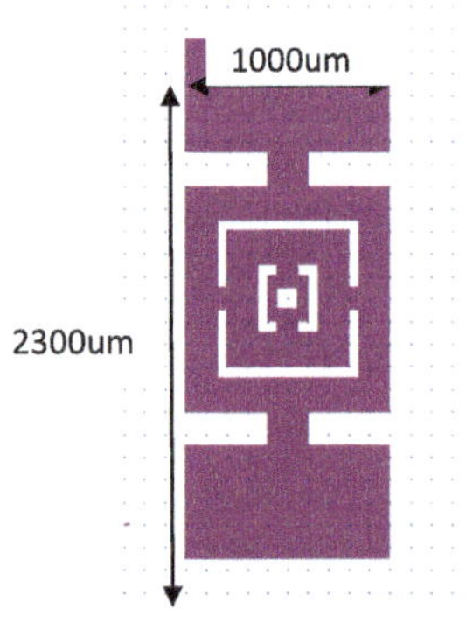

Fig. 6.21 Hotplate design

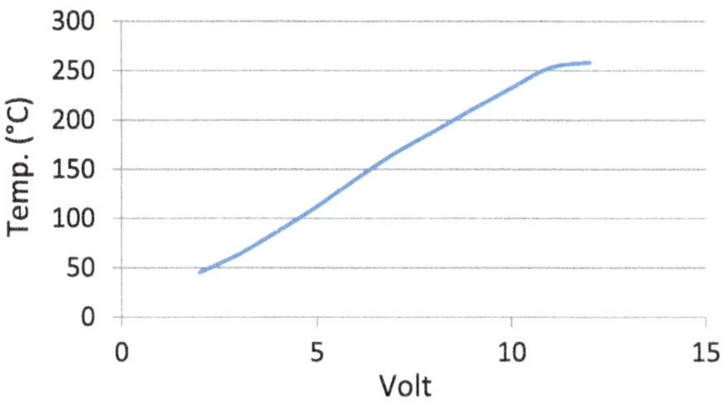

Fig. 6.22 Temperature variation with applied voltage

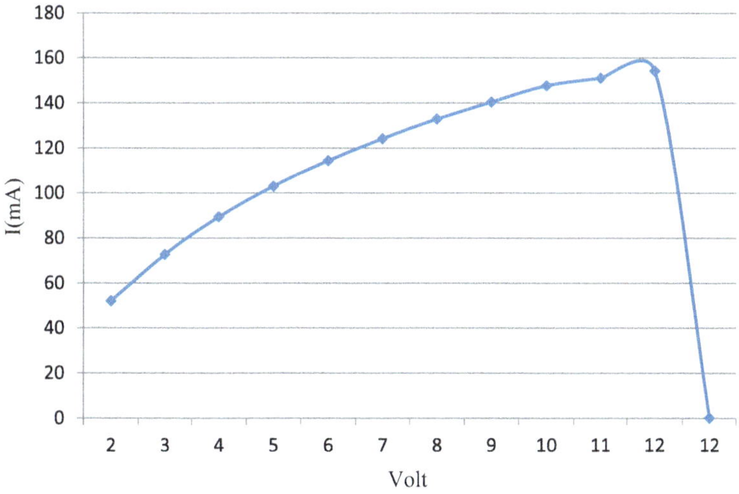

Fig. 6.23 Hot plate current variation with applied voltage

Fig. 6.24 Design of microheater

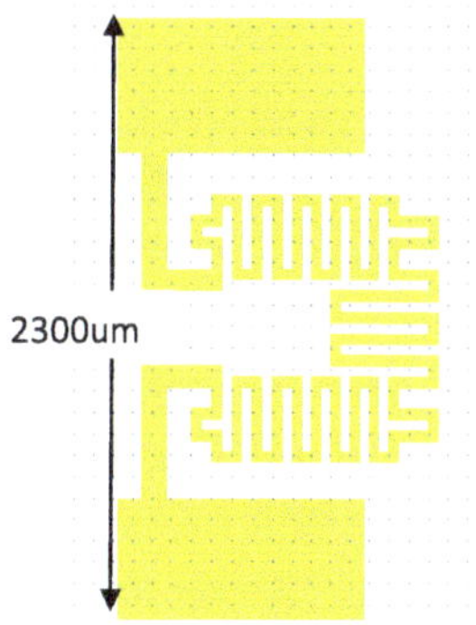

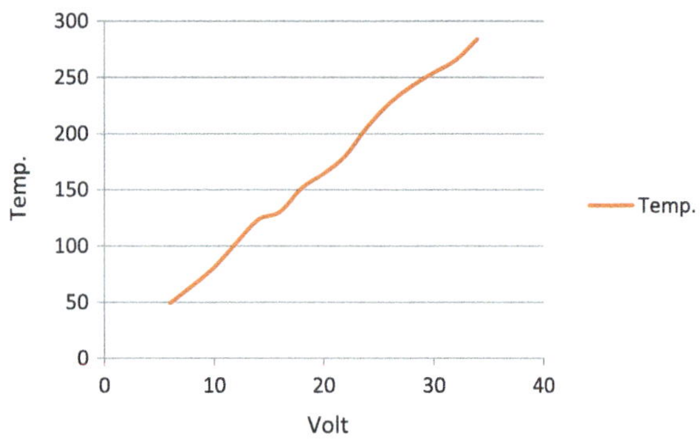

Fig. 6.25 Temp. versus applied voltage

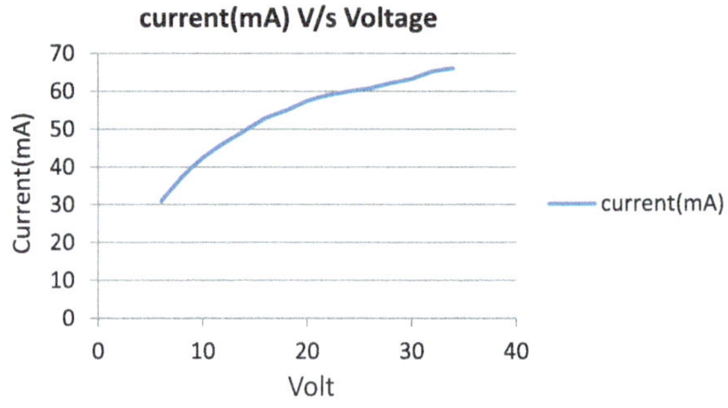

Fig. 6.26 Current variation with applied voltage

This microheater takes ~ 150 mS before breakdown. This can be useful in pyro igniter but higher thermal mass can be a problem for this microheater.

4. This is another serpentine shape microheater with resistance ~ 150 Ω. This micro-hearer was able to attain maximum 243 °C at 27 V. Current behavior was also captured for this microheater using oscilloscope. Plots are given here (Figs. 6.30, 6.31, 6.32, 6.33, 6.34 and 6.35).

It can be concluded from above four microheaters testing that every microheater takes time to settle down. And it is normal phenomena in electronic devices.

6.2.6 Microheater with Back Side Bulk Micromachining of Silicon

RH microheater was fabricated on a DSP wafer. The wafer was bulk micro machined 340 μm from back side using a 2 mm * 2 mm mask. Layers thickness is given here:

a. Substrate: DSP Si-wafer 550 μm.

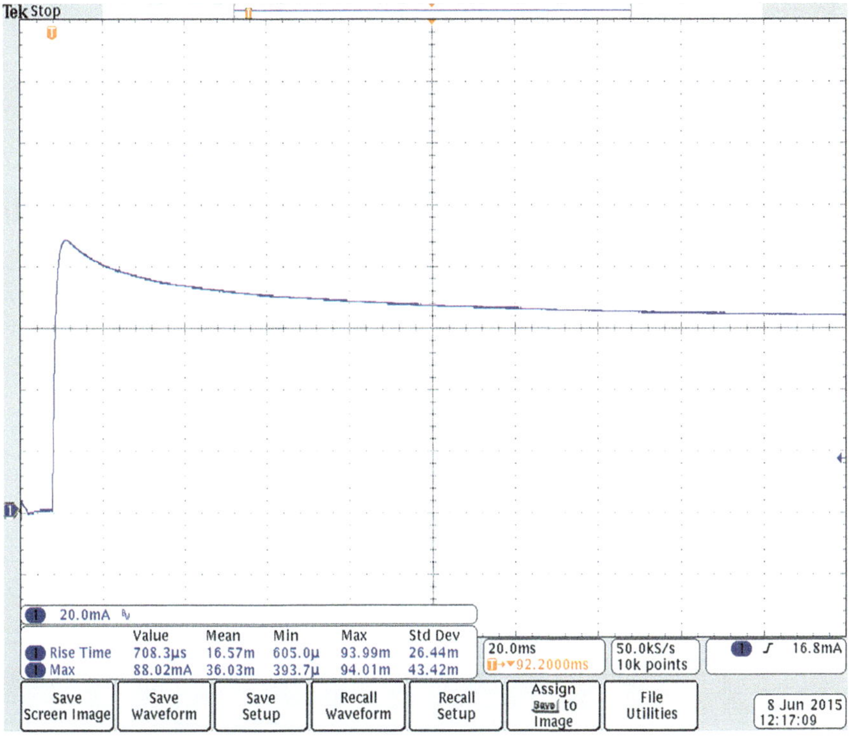

Fig. 6.27 Current behavior at 10 V

b. PECVD Oxide: 0.3 μm(heater side) and 3 μm(back side).
c. Titanium: 0.58 μm.
d. Cavity depth: 340 μm.
e. Diaphragm thickness: 210 μm.

Two dies (1 without bulk micro machine and 1 with bulk micromachining) were packaged on T08 and standard sensors were mounted to measure substrate temperature. The aim was to make a comparison between bulk micro machined and without bulk micro machined microheater (Figs. 6.36 and 6.37).

Above microheater testing results are plotted here for bulk micro machined and without bulk micro machined dies.

It was concluded that substrate temperature is almost same for bulk micro machined and without bulk micro machined dies.

Simulated results are given here for micro machined (cavity depth 340um) and without micro machined die (Fig. 6.38).

There is very less different between micro machined and without micro machined dies because membrane thickness is too high (210 μm). Simulation results were carried out for less membrane thickness (25 μm), it gives very high substrate temperature.

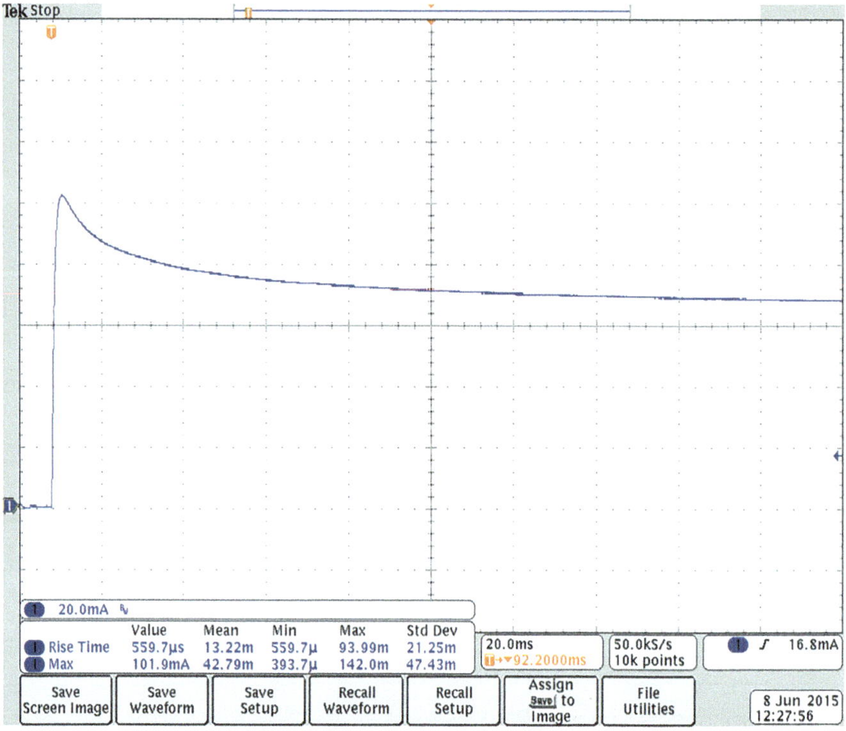

Fig. 6.28 Current behavior at 20 V

6.3 Fabrication Process

After optimizing the design, the fabrication of the structure is carried out on silicon as well as on glass substrate. For Si-wafer, before doing the actual fabrication of heater, thermal oxide is grown and low stress LPCVD nitride is deposited. Finally, the metallization is carried out in E-beam evaporation and it is patterned using liftoff lithography. It is a single mask process.

The glass material is chosen for maintaining both thermal isolation between sensing element and housing assembly, along with provision of mechanical stability. In case of glass substrate, direct metallization process is carried out along with some seed layer for better adhesion purpose. The overall heater foot-print size is $500 \times 500\ \mu m^2$.

Table 6.9 summarizes a comparative study between heater structure developed over silicon and glass substrates. Thermal conductivity in glass case is quite small compared to silicon, hence it is suitable for ignition applications where higher heat is generated:

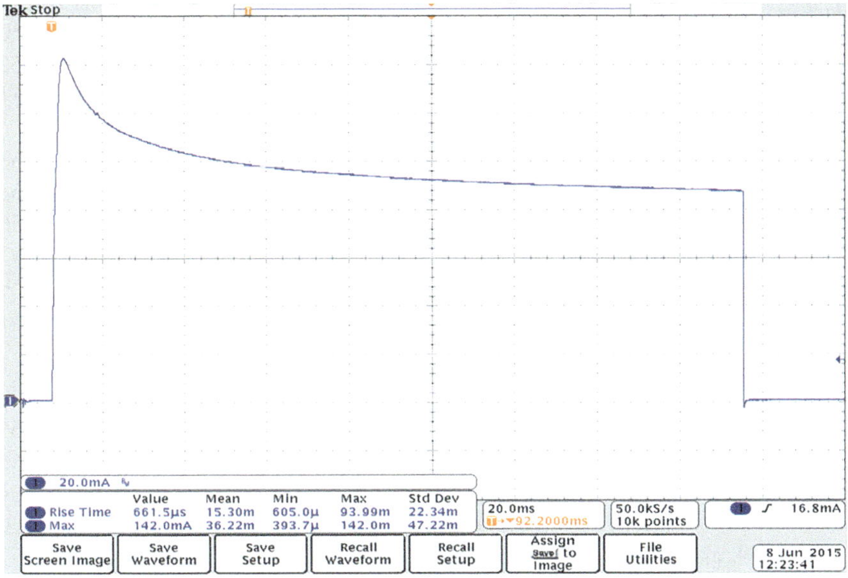

Fig. 6.29 Current behavior at 50 V

Fig. 6.30 Microheater design

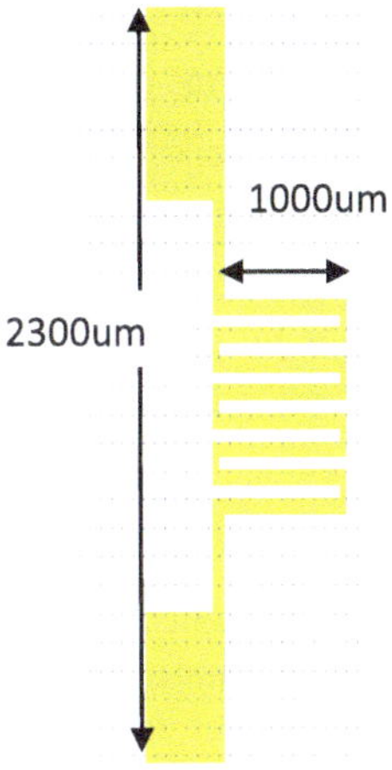

In this development activity, the heating element is realized with a stack of Titanium and Platinum (Ti /P_t), realized by E-beam evaporation technique. The metallization thickness was 0.2 ± 0.1 µm. Ti-layer is used here as adhesion promoting layer [5]. P_t is chosen based on several criteria, such as ease of processing, avoidance of doping as compared to other materials, no tendency of oxide formulation, long-term reliable material, fast thermal response, corrosion resistant, and above all well-established fabrication process. Properties of various heating elements are summarized in Table 6.10.

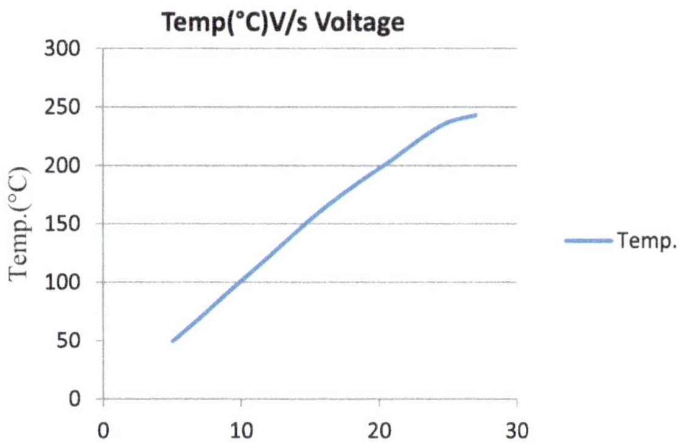

Fig. 6.31 Temp. versus applied voltage

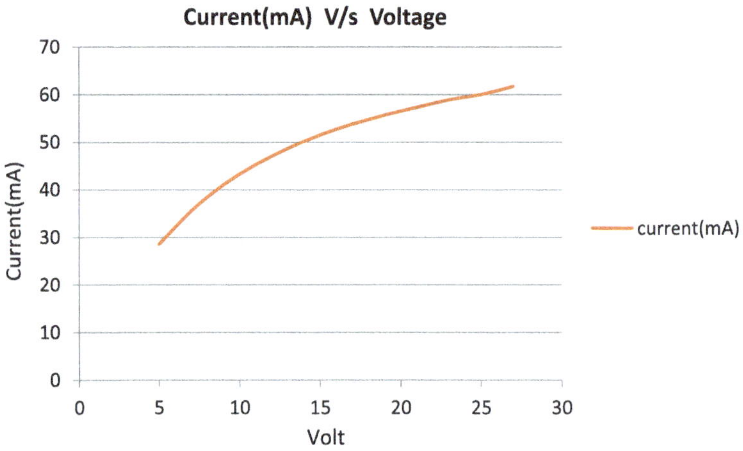

Fig. 6.32 Current variation with applied voltage

6.4 Assembly and Packaging

The microheater structure is realized with multiple wafer bond technique. On the silicon wafer, nozzle, micro-chamber and pad opening were carried out and on the glass wafer the heating element is developed. Dicing mechanism is adopted to open out the front-end part of the nozzle structure and manually alignment is done very carefully for bonding two wafers (Si and glass) [6–8]. A slight mismatch in alignment may bring misleading device operation or failure of the whole heater/igniter assembly. Figure 6.39 shows various parts of heater/igniter assembly.

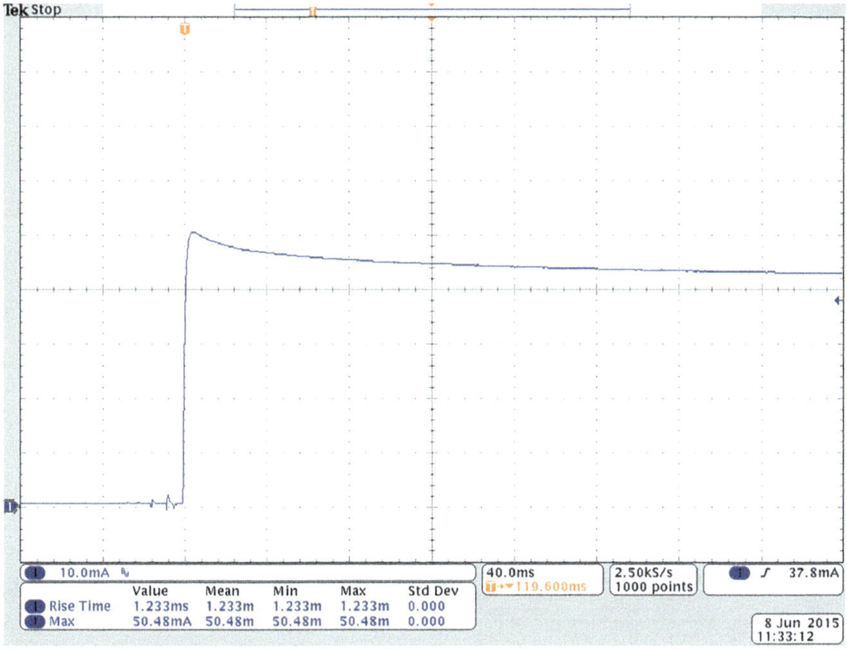

Fig. 6.33 Current behavior at 10 V

Using wire-bond technique, making electrical connection is trivial on the thin platinum pad. Hence, it demands modification of process, finding out the compatible thickness which will support reliable bonding. The prime challenge is to embed hanging wire bond. Laser dicing process is implemented for this purpose.

It is to form parallel groove structure on silicon. Bond wires are inserted in these grooves to facilitate the top cover sealing of the assembly and further to fully protect the wire grooves are filled with conductive epoxies. The complete assembly is shown in Fig. 6.40 and it is attached with the TO-8 header (package), which is a proven assembly for sustaining very high temperature.

6.5 Measured Results

The complete microheater assembly is shown in Fig. 6.41 and the measured I-V curve (current vs. voltage) and electro-thermal parameters are shown in subsequent figures.

From Fig. 6.42, it is clear that the maximum temperature of 220° C is attained with 55 V input and after that, the structure got breakdown with maximum current consumption of around 35 mA.

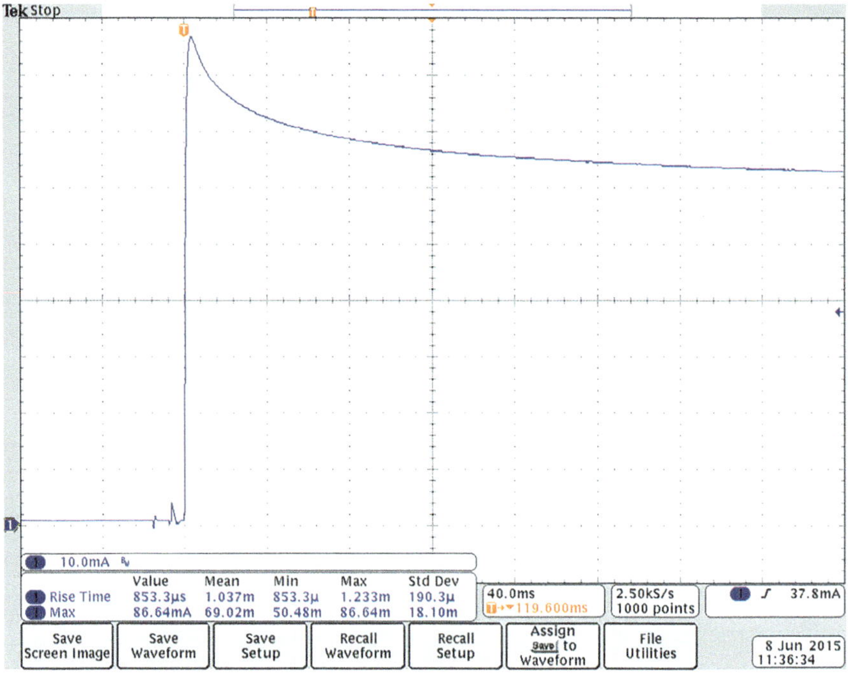

Fig. 6.34 Current behavior at 20 V

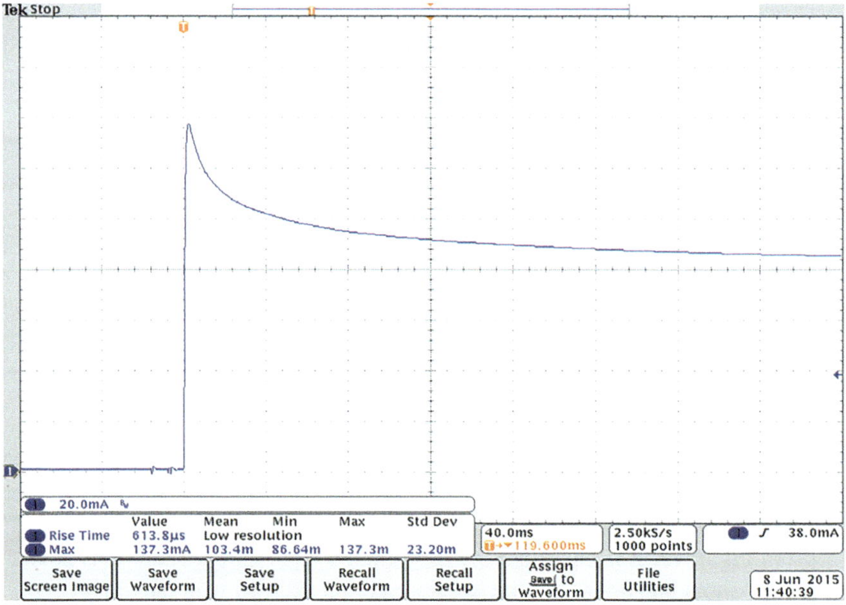

Fig. 6.35 Current behavior at 50 V

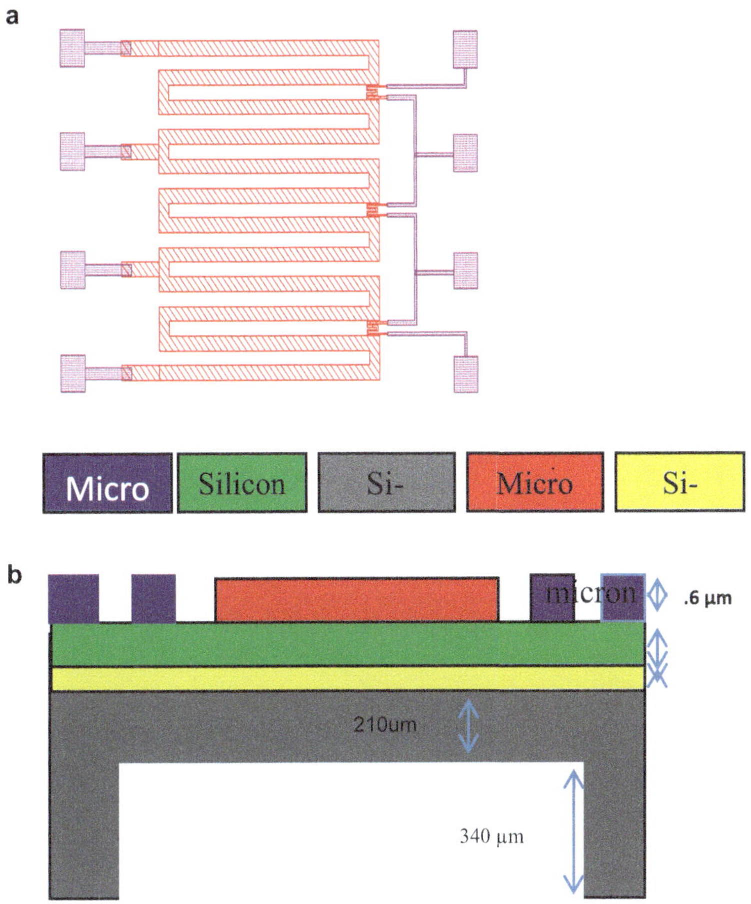

Fig. 6.36 **a** RH sensor microheater design. **b** Block diagram of BM microheater

Thickness of the substrate plays an important role in the rise in temperature, as it contributes in determination of thermal mars. Thermal imaging of the structure is also captured and depicted in Fig. 6.43, which deciphers the uniform pattern of heat energy (Fig 6.44).

The measured results of the realized heater are tabulated in Table 6.11.

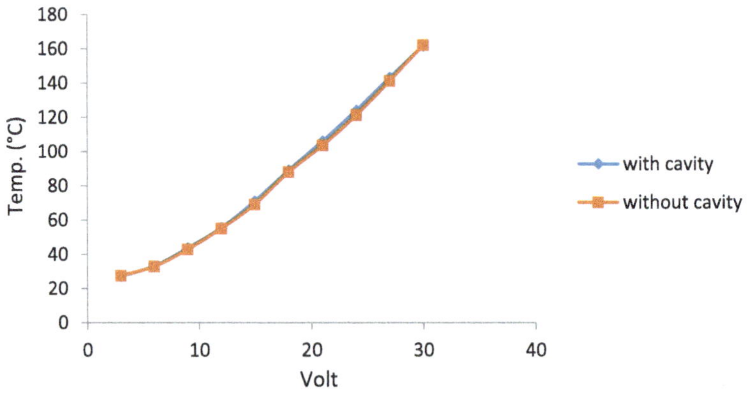

Fig. 6.37 Temp. variation with applied voltage for BM and without BM process. [BM-Bulk micromachining]

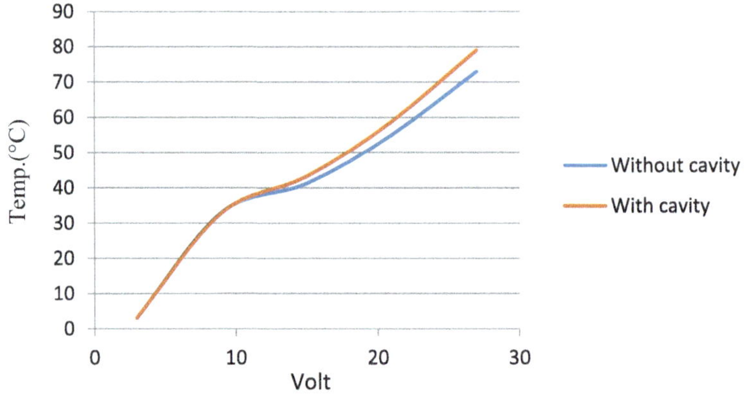

Fig. 6.38 Simulated results for BM and without BM microheater

Table 6.9 Parameters comparison on silicon and glass substrate

Parameters	Silicon	Pyrex glass
Thermal conductivity (W/mK)	140	1.18
Melting point (°C)	1414	1700
Elastic modulus (GPa)	112.4	64

Table 6.10 Properties of various metal layers

Properties	Titanium	Platinum	Poly silicon
Thermal conductivity (W/cm/$^\circ$C)	0.219	0.716	0.148
Electrical resistivity (Ω.cm)	4.8e−5	1.06e−5	82
Thermal expansion coeff. (μm/m.K)	8.6	8.8	4.7
Specific heat (J/g/$^\circ$C)	0.523	0.133	10^5
Density (g/cm^3)	4.51	21.45	2.32

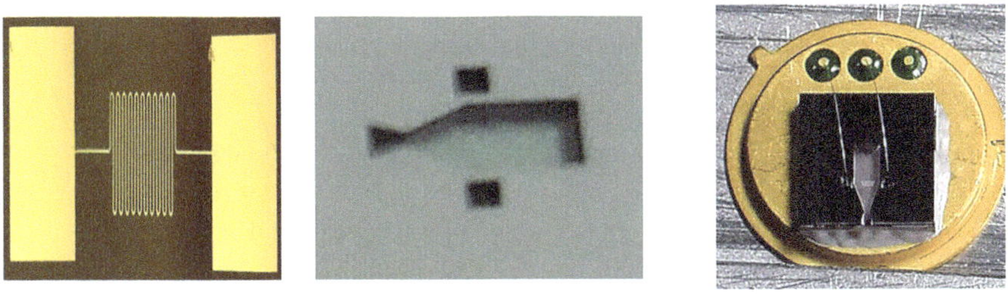

(a) (b) (c)

Fig. 6.39 **a** Layout of the realized spiral shape microheater **b** chamber after etching **c** heater final assembly

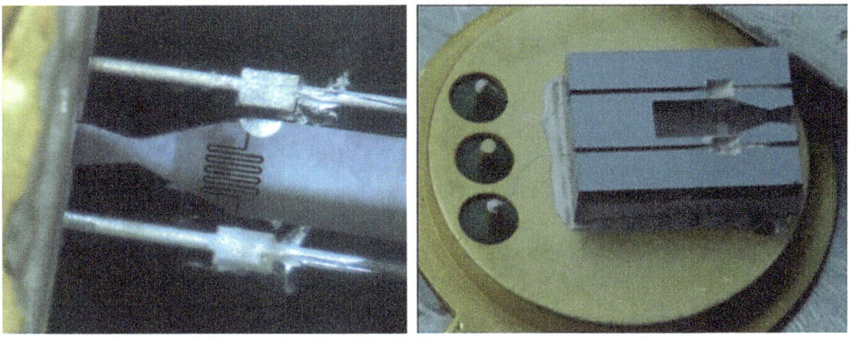

Fig. 6.40 Wire bond inside the silicon tunnel

Fig. 6.41 Assembled heater on thin outline package

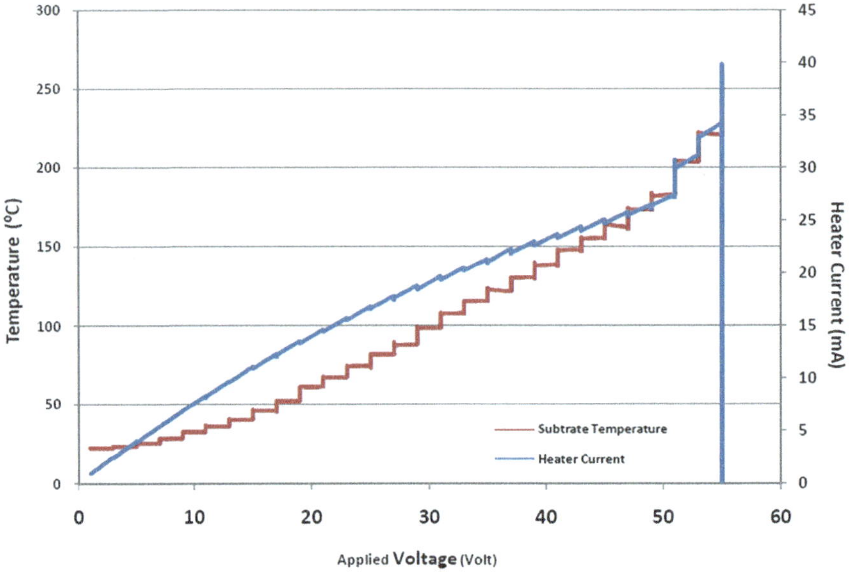

Fig. 6.42 Measured results of meander shaped microheater on silicon substrate

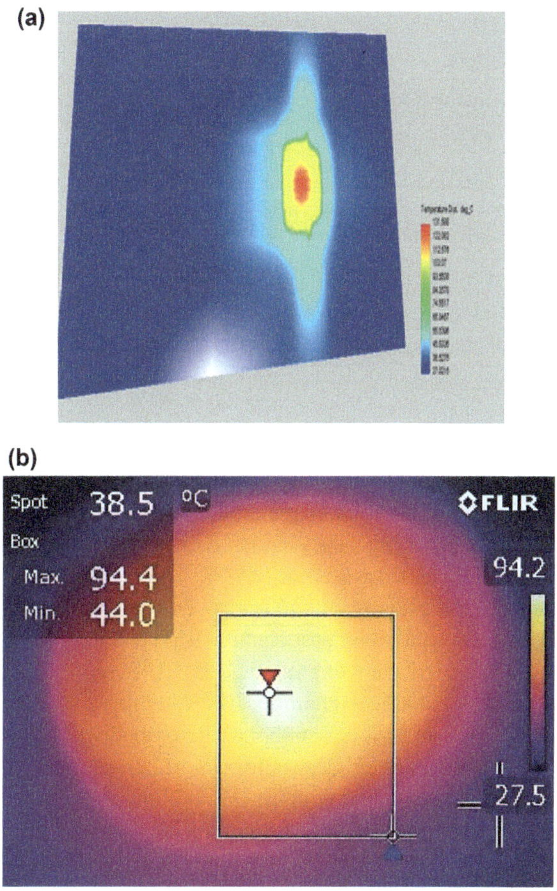

Fig. 6.43 Thermal image of the heater on glass substrate **a** simulated **b** measured

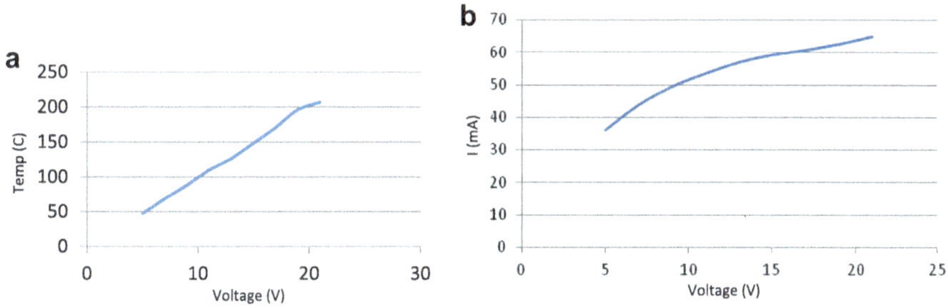

Fig. 6.44 Measured plots of **a** Temp versus voltage **b** Current (mA) versus voltage

Table 6.11 Tabulation of the measured results

Parameters	Measured values (Glass)	Measured values (Silicon)
Excitation voltage	~ 21 V	~ 55 V
Attained temperature	207 °C	~ 210 °C
Resistance	100 ± 20 Ω	1200 ± 50 Ω
Current (max.)	150 mA	55 mA
Settling time	40–60 (ms)	40–60 (ms)
Rise time	633(μs)	633(μs)
TCR (Ω/Ω/°C)	0.0025	0.0018
Igniter efficiency (%)	32	5

6.6 Summarization

Meander shaped heater realized on glass and silicon substrate is presented in this article. Assembly and characterization aspects are covered in detail and further comparative analysis is carried out. The present approach shows that heater topology realized on glass results in improved efficiency for igniter application due to lower thermal conductivity associated with the wafer. Present topology is fabricated using micro-fabrication technique and DRIE-based etching process is carried out for the realization of silicon chamber. The assembly of chamber with the heater is carried out manually and further wire bonds are embedded by having tunnel option in the silicon wafer so as to avoid destructive interferences while testing and sealing. Further this assembly is attached with the thin outline package for characterization. This concept of planar realization will eventually be useful for pyro igniter application for generating small thrust in array configuration. Present approach is based on heater fabrication using IC technique and further performance enhancement can be achieved by carrying out micromachining which is already standardized for silicon process.

Questions

1. What is microheater?
2. What are the different end usages of microheaters?
3. What are the salient features of microheater design?
4. What do you mean by power compensated microheater design?
5. Discuss on the various types of microheater designing in MEMS/microfabrication process?
6. What is pyro igniter?

References

1. Roy S, Sarkar CK (2016) MEMS and nanotechnology for gas sensors. CRC Press
2. Hwang W-J, Shin K-S, Roh J-H, Lee D-S, Choa S-H (2011) Development of micro-heaters with optimized temperature compensation design for gas sensors. Sensors, 2580–2591
3. Singh K (2015) Development of indigenous metal oxide gas sensors with embedded heater using CMOS process. Punjab Acad Sci J 13–14(1 & 2):1–4
4. Baliga BN, Tiwari DN, Singh K, Nagachenchaiah K (2009) Effect of Polymide variation and its curing temp. on CMOS based capacitive humidity sensor and characterization of integrated heater. Sens Transducer J 144–154
5. Chang W-Y, Hsihe Y-S (2016) Multilayer microheater based on glass substrate using MEMS technology. J Microelectronic Eng 149
6. Chung G-S, Choi S-K, Nam H-D (2001) Design, fabrication, and characteristics of microheaters with low consumption power using SDB SOI. In: Proceedings of SPIE - the international society for optical engineering, vol 4601
7. Toskov S, Glatz R, Miskovic G, Radosavljevic G (2013) Modeling and fabrication of Pt micro–heaters built on alumina substrate. In: Proceedings of the 36th international spring seminar on electronics technology, pp 47–52
8. Anjitha RG, Ahirwar S, Karthikeyan L, Subhash J, Singh K, Andhiwal A, Kumar Basu P: Design, fabrication and packaging of an optothermally activated nanocrystalline Pd-Zno based selective CO sensor on a screen printed in-plane heater. ACS Appl Electron Mater J

Chapter 7
Critical Aspects in Micro-devices

Realizing the micro-devices is a challenging task to the practical engineers involved in this specialized field. Not only the design and device fabrication play an important role in this holistic activity, but also packaging or post-fabrication activity is considered as one of the key process for getting success in this domain. As compared to standard micro-fabrication or CMOS process, MEMS technology differs entirely because of its several unique features, engaged at the every process steps of its realization starting from scratch to the end product. This particular chapter highlights such three important critical aspects of micro-device realization. Those are: wafer bonding, packaging aspects in terms of HMC and bond wire effect. Critically every section has been touched upon here with practical/experimental results obtained by the authors group.

7.1 Wafer Bonding in MEMS Technology

MEMS technology has some unique features as compared to standard micro-fabrication process. One such unique feature is capability of bonding of wafers. With this method, multiple wafers can be bonded together. This bonding can be done in three main ways viz. anodic, eutectic, and fusion bonding. Although, in some literature one can find 'Glass-frit' bonding category, which is very rarely used process and little bit dirty process in context with cleanliness of the wafer or contaminants. Out of the three aforesaid methods, the first two categories are considered to be much popular and widely used in micro-machining method. The third category is a very high temperature (~1050 °C) process needs a process chamber and strict control. Almost all the MEMS sensors employ either eutectic or anodic bonding process. Both of the methods have their own merits and demerits. Depending upon the application area or device deployment, one can use either of the two. Cleanliness of the wafer surface plays an important role in this bonding method. A small particle

K. Singh and A. Karmakar, *Micro Electro Mechanical System (MEMS) Technology for Multifaceted Applications*, https://doi.org/10.1007/978-981-96-6162-6_7

can create a crack or void in the bonded wafer structure. Hence, before initiating the actual bonding process, various surface cleaning methodologies are adopted in MEMS industries. Usually, SC1, SC2, RCA cleaning methods are implemented to remove the organic or inorganic contaminants. Further, ultrasonic or mega-sonic cleaning methods are applied to remove any loose particles over the surface of wafer. After, the cleaning, rinse and dry method is adopted to prepare the wafers ready for wafer bonding [1–3].

Anodic bonding is implemented to bond one silicon wafer to one glass wafer at a temperature around 450 °C in an inert ambience. The glass wafer should be sodium-rich. Pyrex-7740 or borosilicate glass is example of sodium-rich glass. The actual bonding mechanism is – 'anodic' type with the application of external electric field. Further, the coefficient of temperature expansion (CTE) of both materials should match well at the bonding temp. A certain pressure is maintained inside the process chamber in a N_2-gas ambience. Proper thermal profile (ramp up or down nature) is maintained to produce reliable wafer bonding. This method becomes very useful in pressure sensor, acoustic sensor, micro-machined antenna, aperture-coupled antenna, etc. devices or sensors. The bond strength of this process is proven to be very high, yielding a very reliable process [4–6].

Whereas, in the eutectic bonding method, a sandwich gold layer is introduced between two silicon wafers. At a suitable temperature called 'Eutectic Temperature,' fusion or melting of this sandwich gold layer happens and at ~ 363 °C eutectic bond is formed between silicon and gold. In case of realizing cavity type μ-wave filter, or bonding of multiple wafers for vertical integration, this method seems to be very reliable. One should use judiciously the right category of wafer bonding depending upon its practical usage [7].

In the next subsection, we will discuss more about 'Anodic bonding' process details.

7.2 Anodic Bonding Process

It is an electric field assisted wafer bonding process to seal glass to either silicon or metal without introducing an intermediate layer. In general, this bonding mechanism is widely adopted to seal glass to silicon wafer in electronics and micro-fluidics. Sometimes, it is known as 'electrostatic bonding'. The stringent requirement for this process is the ultra-clean surfaces of silicon/glass wafers. The coefficient of thermal expansion (CTE) value of the processed glass needs to be similar to those of the bonding partner. The bonding can be done at temperature of 250–450 °C. The glass wafers needs to be highly sodium ion (Na^+) rich. The procedural steps of anodic bonding are as follows:

1. Contact substrates.
2. Heating up substrate.
3. Bonding with the application of E-field.
4. Cooling down (or annealing) of wafer stack.

The whole process is characterized by following variables:

1. Bond voltage.
2. Bonding temperature.
3. Current limitation.

As already, we have given an outline on various bonding mechanisms implemented in micro-machining process. In Table 7.1, the comparison among various bonding mechanisms have been chalked out.

The anodic bonding procedure is able to bond hydrophilic or hydrophobic silicon surface equally effectively. Roughness of the surface should be less than 10 nm and free of contaminations on the surface for the procedure to work properly.

Generic process tool set-up is shown in Fig. 7.1. The top tool works as a cathode and the chuck as anode. Ion drifting in bond glass is influenced by the applied E-field. The whole process can be step-wise divided in two ways:

(a) Formation of depletion region through drifting of mobile sodium ions (Na^+).
(b) Drifting of oxygen ions in the depletion region.

Difference in TCE parameter of silicon and glass with temperature variation is depicted in Table 7.2.

Table 7.1 Comparison of various bond mechanisms

Parameters	Anodic	Fusion	Eutectic	Solder
Bond strength	Very strong	Excellent	Strong	Strong
Hermiticity	Good	Excellent	Excellent	Excellent
Temperature(°C)	250–400	200–1000	> 363	57–400
TCE mismatch	OK	Good	OK	Poor
CMOS compatibility	OK	Poor	OK	Good
Planarization capability	Poor	Very Poor	Good	Good

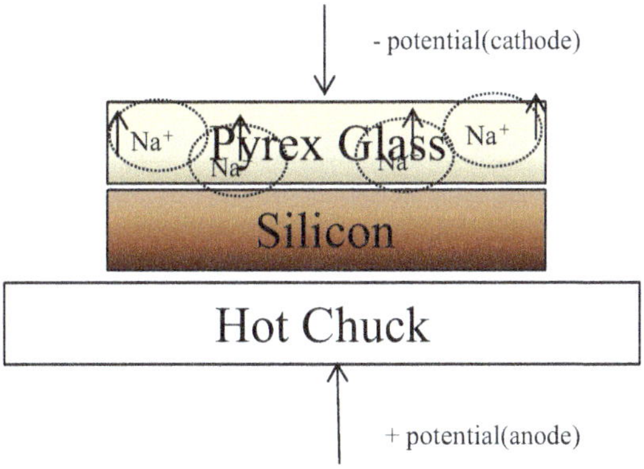

Fig. 7.1 Generic experimental set-up for wafer bonding (anodic)

Table 7.2 Silicon and Pyrex 7740 characteristics at various temperatures

Temperature (°C)	Silicon TCE (ppm/°C)	Glass TCE (ppm/°C)
300	3.78	3.3
350	3.86	3.32
400	3.96	3.40
450	4.02	3.60

In fact, with the elevation of chuck temperature along with high potential basically accelerate the mobility of Na^+ toward the anode and thus it forms the space charge in the glass substrate. Overall experimental set-up can be electrically modeled as a combination of 'R' and variable 'C' as shown in Fig. 7.2.

Here, the variable 'C' denotes the capacitance effect due to depletion region. The variability of the capacitance value depends on the movement of mobile Na^+ ions. The resistance (R) is offered by the path (glass substrate). If we observe the process recipe, we can observe that the current density increases with the formation of depletion region, and afterward it decays exponentially to a minimum value.

Table 7.3 demonstrates the typical process parameters and conditions of a wafer bonding (anodic) method. Each of every parameter has a great impact on the quality of the bonding process. After the actual bonding process, slow cooling over several modules has to take place. It is supported by purging with inert gas, especially nitrogen (N_2) ambience. The cooling time depends on the difference of CTE for the bonded materials: the higher the mismatch of CTE values of materials, the longer cooling time is required. In Table 7.4, the effect of various process parameters has been shown in a symbolic way. The lower chuck electrode material also plays an important role in the bonding process, as it may produce unwanted contamination particles. Multiple time uses of same chuck are generally avoided for better process yield.

As pyrex glass constituent comprises of Na, B, Al atoms so mathematically this process can be shown as

$$\begin{aligned}
&[\mathrm{AlO_4}]\mathrm{Na} + \mathrm{H}^+ \leftrightarrow [\mathrm{AlO_4}]\mathrm{H} + \mathrm{Na}^+ \\
&[BO_4]\mathrm{Na} + \mathrm{H}^+ \leftrightarrow [\mathrm{BO_4}]\mathrm{H} + \mathrm{Na}^+ \\
&\mathrm{Si} + 4[\mathrm{OH}]^- \rightarrow \mathrm{Si(OH)_4} + 4\mathrm{e}^- \\
&\mathrm{Si(OH)_4} \rightarrow \mathrm{SiO_2} + 2\mathrm{H_2O}
\end{aligned} \tag{7.1}$$

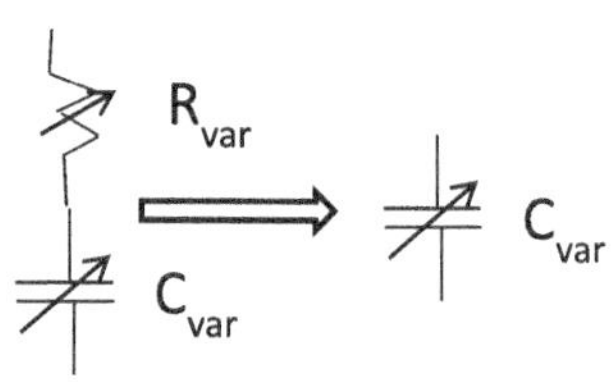

Fig. 7.2 Equivalent model of the bonded assembly. **a** Initial. **b** After bond formation

Table 7.3 Process criteria for bonding operations

Process parameters	Process requirements
Electric potential (~1000–1200 V)	Glass wafer (~20 mils)
Chuck temperature(~400 °C)	Electrode material (Graphite)
Current density(1A/m^2)	Surface roughness(<1um)
Atmosphere (~1e5 bar)	SiO_2 thickness(<2000 Å)
Process time (30–60 min after stabilization)	ΔTCE between Si and Glass(min)

Table 7.4 Effect of process parameters

Process time	Temperature	Pressure
↑(Bonding)	↓	–
↑(Bonding)	–	↓
↑ (Ring width/newton fringes)	↓	–
↓(Ring width/newton fringes)	–	↓

Table 7.3 summarizes the process parameters and process requirements of the bonding process.

7.3 Generic Process Flow of a Standard and SOI Type Silicon Wafer for Wafer Bonding

Figure 7.3 demonstrates a wafer bonding process for standard silicon wafer. As a prerequisite, the bonding surfaces should be thoroughly cleaned to avoid any unwanted contaminations or particles, which may create further void or crack in the bonded wafer.

In general, for making multiple cavity or trench the base wafer is processed either in wet etchant or implementing deep reactive ion etching (DRIE) process. For precise control in this cavity or trench opening, usually hard masks made of thick oxide or nitride layer is used. For the bonding process, this hard mask should be removed in later time. Rigorous chemical treatment or cleaning isn't viable for the bonding process, as already the front surfaces may contain main sensing parts or electronic parts, which may be corrode with chemical treatment. Other constraints in this unique MEMS process are:

1. Surface area or area of bonding is very minimum.
2. Handling of wafer is very important, as it becomes fragile in most of the time.
3. Wafer may break if wrongly handled or a forcefully water jet is applied during dicing process.
4. Drying of bonded wafer is also very critical. Extensive care is taken before and after the bonding.

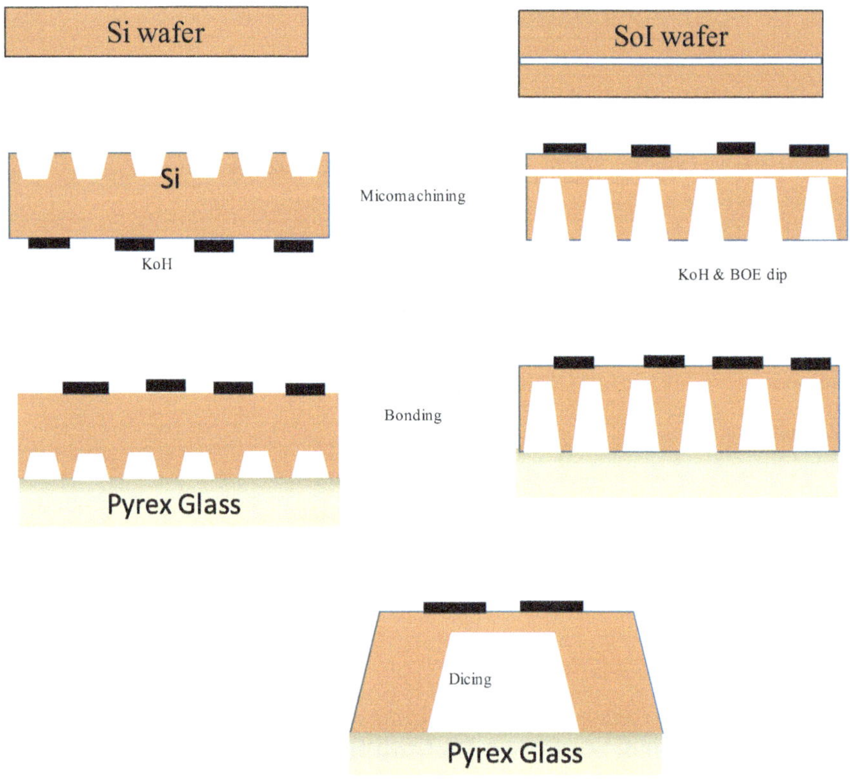

Fig. 7.3 Bonding process flow in case of standard and SOI-wafer

Initially, the quality of wafer bonding process is ensured by the visual inspection (VI) method. Formation of bubble inside it or the observation of any crack or noticing charred portion (blackish) are some clear cut indication of poor bonding. In general, immediately after the wafer bonding process, the wafer is kept for some time in clean room ambience to attain the room temperature, followed by dipping it into De ionized (DI) water. Seepage of water in the bonded structure is a clear indication of failure of bonding mechanism. If it is passed through this process, further it is sent to IC-assembly or packing area for dicing purpose. Easier separation of bonding partners indicates the poor quality of the process. On the contrarily, hard separation or breakage of some silicon portion infers a good quality bonding mechanism. Further, the quality of the bonding process is evaluated quantitatively with the aid of 'Bond strength,' which is expressed in kilo-gram-force (kgf). Either 'Die-Shear' tests or 'Bond-Pull' tests are conducted in this context.

The next section will discuss about some experimental results conducted by authors group to qualify the wafer bonding process [1–3].

7.4 Experimental Results

To inspect the quality of wafer bonding process, generally two methods are there: non-destructive and destructive. Non-destructive tests include thorough visual inspection, fine and gross leak testing, water bubble tastings, etc. These kinds of tests can predict qualitatively the bond quality. However, to measure the bond quality one should adopt destructive test only. Bond-pull test, die-shear test, TEM, etc. are used traditionally for quantifying bond strength.

In the present section, we will discuss above such die-shear test results of dies conducted with the standardization of MIL-STD-883. Standard 550 μm silicon wafer and 500 μm thick Pyrex-7740 glass wafer bonding was done. The cleaning methodology also impacts the quality of bonding, as summarized in Table 7.5.

From the above data, it can be inferred that cleaning with DI-water before and after bonding process is comparatively less effective than that of treatment in DI-water followed by UV-cleaning at 50 °C. The UV-frequency was set at 34 kHz. In a single wafer, usually at 5 different locations, the bond strength is measured by picking up different dies. Further study was carried out to chalk out information on the surface-area of bonding allowed in three different occasions. Table 7.6 briefly outlines the fact.

The above experimental outcome indicates that, weaker bond strength is noticed in case of SOI-wafer as compared to standard Si-wafer.

Table 7.5 Bond strength characterization with different cleaning methodology

Bond strength(~kgf) DI cleaned (1st Method)	Bond strength(~kgf) DI &UV treated wafer(2nd Method)
15.00	36.50
17.60	31.20
21.80	25.00
21.50	10.90
6.50	35.00

Table 7.6 Bond strength comparison of cavity wafer

Bond strength (kgf) plain wafer	Bond strength (kgf) cavity std wafer	Bond strength (kgf) Cavity SOI-wafer
34.00	21.20	12.70
37.50	38.20	17.50
30.00	20.90	13.50
28.50	16.90	16.55
7.50	17.50	5.80

7.5 Packaging Aspects

This section discusses about the packaging concepts in hybrid microelectronic packages. Hermetic seam sealing is the vital and final assembly operation in the fabrication of Hybrid Micro Circuits. HMC is vacuum baked prior to sealing and sealing process is carried out in the nitrogen purged dry Glove box chamber. Lid alignment is most challenging and critical in sealing process which in turn depends on smoothness of sealing flange, rough surface leads misalignment of lid and to unsuccessful leak test. Delidding a hermetically sealed package with smooth surface on sealing flange is the requirement of successful resealing. Presently dressing activities involves removing burs, smoothening the surface are carried out after delidding process, which is necessary to achieve smooth surface around the sealing flange in order to carry out resealing process. Dressing operation introduces conductive particle inside the cavity of the package. Elimination of dressing operation is challenging and the need, to avoid particles entering inside the cavity. Delidding techniques and process parameters have been developed which minimizes the required dressing operation, which have improved the performance of the resealed HMCs in PIND test. The main consideration in HMC is to have cost effectiveness for which package cost contributes 20–30% of the overall cost. This article shows the methodology of delidding and resealing of package which is successfully implemented in the standard HMC line.

7.5.1 Sealing and Delidding Process

Seam welding, also called seam sealing or resistance welding is the most popular and widely used process for hermetic sealing of HMC packages [6].

In seam sealing, the lid is positioned on the package, two truncated cone shaped copper alloy electrodes are allowed to travel along the top edge surface of the opposite sides of the package lid. These electrodes deliver high current pulses that flow through the lid-package interface, laterally across the lid as well as through the package and out the other parallel electrode. The travel rate of electrodes, the intensity and duration of current pulses, and the interval between pulses are controlled to provide the optimum molten region of overlap with the minimum energy input [7]. Seam sealing localizes the heat to the point of contact of the lid and electrodes, focusing sufficient power into the small seal area so that the weld is completed before the rest of the package have time to heat up [8–10]. The temperature at the point of contact is about 1500 °C and has been controlled by the current pulse amplitude. Figure 7.4 shows the set-up of sealing/resealing of HMC packages.

In the delidding process, operator places the package face down on the machine's slide table, and pushes the package into the reference corner. One edge of the package, e.g., a long side, will be overhanging the edge of table and will be machined momentarily. With the package face down in this corner, the machine is able to make an

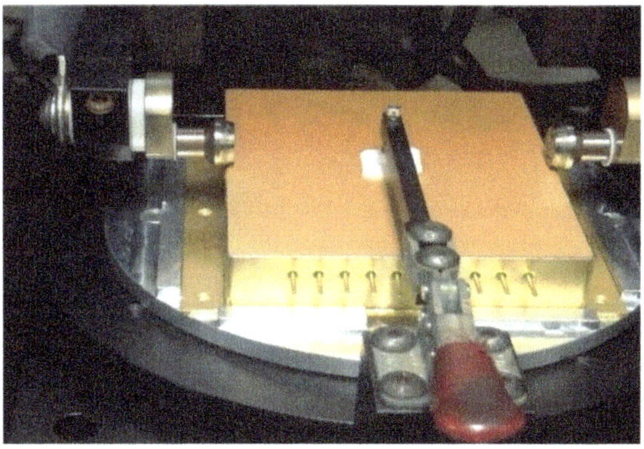

Fig. 7.4 Sealing/resealing of HMC package

accurate cut by referencing the top of the package and two adjacent walls. The operator rotates the package by 180 degrees, clamps the package again, and starts the table. When the second-long side has been machined, the operator must make one adjustment before continuing with the two remaining short sides; the cut length knob must be adjusted for their length. If the package is square, this step is not necessary. Once all four sides have been machined, the package is removed from under the clamp, the package is lifted up, and the lid will remain behind on the table. If the lid is still attached to the package, a tapping on the package gently will cause the lid to fall free. At this point, the package physically can be resealed without dressing or cleaning the flange.

Delidding of welded packages by precision sawing is used and is the most recommended method. Optimum process parameters and acceptable tolerances were established for larger packages for the reason more glass to metal seals which are typically the weakest part of the package and complex circuits are chosen. Since a method developed for the larger package would apply to the smaller one, $2'' \times 2''$ welded butterfly packages were selected for delidding and resealing process qualification. Step lids were used for resealing the packages. The lids were 15 mils thick in the center and 5 mils thick at the edge. Lid plating was identical to the package plating. All sealing and resealing was performed on a resistance seam welder equipped with dual rolling electrodes.

7.5.2 Challenges in Package Delidding Process

Particle control and achieving smooth surface finish on sealing flange while delidding is the important requirement for successful delidding and resealing for hybrids which must pass particle impact noise detection (PIND) test and seal leak test.

Dressing operations of the seal flange to remove burrs or irregularities after delidding exposes the circuit to particle generation and subject the package and circuit to a very definite potential for damage. Elimination of dressing operation by

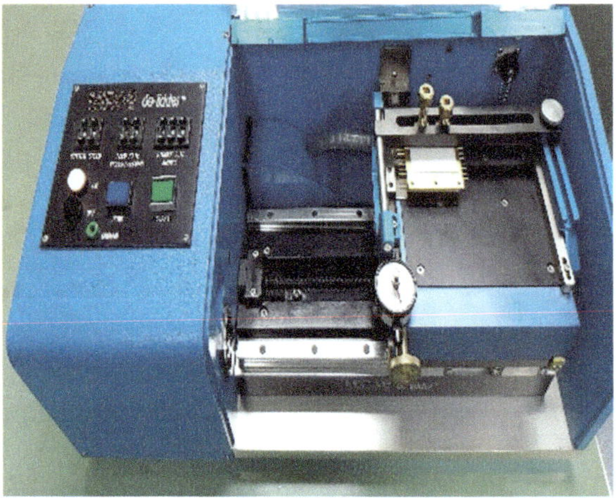

Fig. 7.5 Delidding of HMC package

achieving smooth surface finish while delidding is important in particle control and for successful leak test after resealing.

A critical element in delidding process development is minimizing step height. If the cut along a side of the package is too deep, the blade machines away part of the seal flange, creating a step from the original seal flange down to the machined surface. If the cuts down two adjacent package sides are not quite identical, a step occurs where the two cuts intersect. Too high a step causes several problems for resealing. First, the welding electrodes will try to deform the lid over the step, introducing a bending stress and a potential crack zone along the ridge. Second, the weld energy distribution can be affected, particularly when the step configuration allows the electrode to pass current into both the seal flange edge and the step edge; this creates two marginal weld regions, and a probable leak site.

Delidding approach uses the bottom of the package as the reference plane. Top and bottom of the package each have a planarity tolerance, and are not usually required to meet a specific parallelism dimension. Clamp pressure must be adequate and uniformly distributed to hold the package securely in the reference corner during each machining pass. Excessive pressure, however, tends to distort the package, cracking the glass to metal seals and produces variation in flange step height. Achieving uniform clamp pressure/force distribution during delidding process is challenging in reducing step height. Figure 7.5 shows the set-up for delidding of HMC packages.

7.5.3 Delidding Process Optimization

As shown in Fig. 7.6 an optimized cutting speed, cut width and cut depth will produce smooth surface finish on sealing flange and dressing operation free delidding.

Firstly, the packages have walls 40 mils thick and welds are 20–30mils wide depending on sealing parameters. Cut width to 10mils less than the package wall or

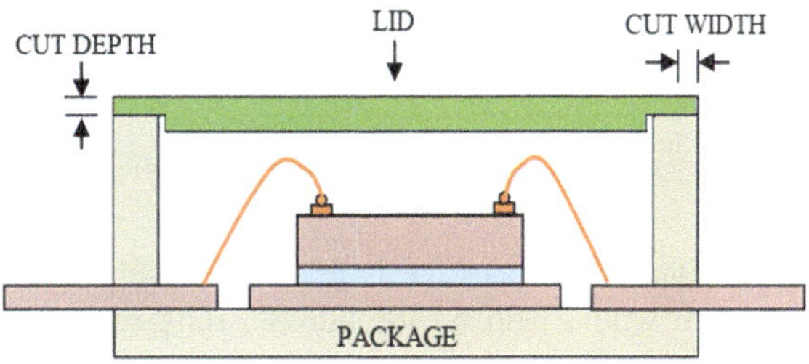

Fig. 7.6 De-lidding width and depth

seal flange is important for particle control. Since the clamp keeps the lid clamped to the remaining portion of the seal flange, providing a physical barrier to particles. Furthermore, by not cutting the whole width of the wall, the lid cannot fall into the package. If the cuts were made too wide and the lid did fall in, not only would the wire bonds be smashed, but the top of the lid would have moved with respect to the seal flange, invalidating the reference plane for depth of cut, this results in subsequent cuts being too deep. Cut width of 30 mils wide removes the lid completely from the package and the cut depth is equal to the thickness of the lid eliminates the dressing operation after delidding. Cut length is to be set precisely adjusted as necessary to prevent particles from getting into the circuit. The emphasis is on making a clean cut, machining away only the weld region and no more, preventing damage to the package header and circuitry, and making sure resealing will be successful (Fig. 7.7).

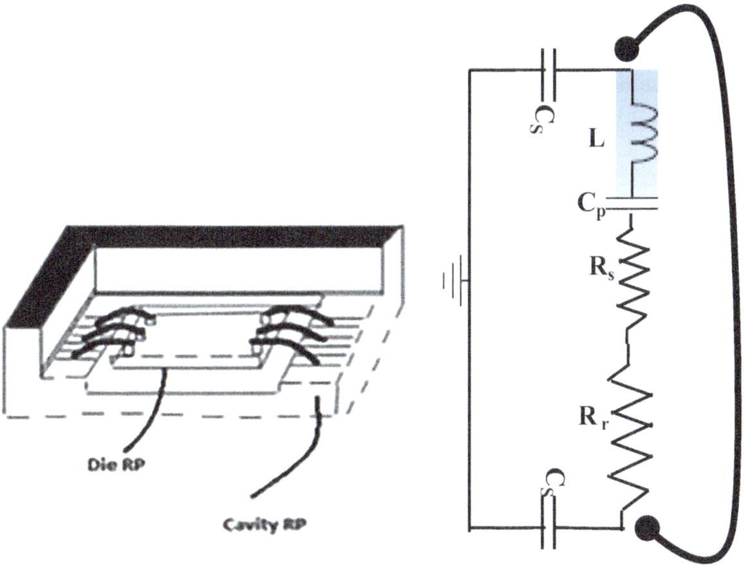

Fig. 7.7 Illustration of bond wire with reference plane and the equivalent model

Table 7.7 Delidding parameter for ceramic and metal packages

Stroke size	Delidding width	Delidding depth	Cutter speed dial setting	Torque force
0.3–0.9 inch	5–40 mil	5–18 mil	10–99	3–4 kg

Secondly, the package is delidded upside down, so any particle entering the package would land on the lid. A strong vacuum pickup is located almost flush with the saw blade, collecting almost all of the particles as they are created.

Thirdly, a Teflon sheet of thickness 5 to 10mil is placed on the package to accommodate the top and bottom package planarity tolerance and optimized Clamp pressure is applied for uniform pressure distribution on package to hold the package securely in the reference corner during each machining pass. Excessive pressure, however, tends to distort the package, cracking the glass to metal seals. It is also required that the clamp pressure should be low enough so that a given package can be clamped not just one time, cut many times.

Finally, an optimized table feed rate, starting with minimum and gradually increasing to the maximum on each side while delidding produces clean and smooth surface finish on seal flange. Starting table feed rate with high, contributes to excessive blade loading, blade can't machine away the metal fast enough and the blade stalls out, vibrates, chatters, and makes rougher cut than desirable.

The optimum values of delidding parameter were determined and the value selected for each of the key variables is presented in Table 7.7.

Thorough cleaning is recommended prior to sealing in order to remove particles and contamination that may have entered the package during the rework operations. Resealing is performed using a new lid and using the identical equipment and welder settings as for normal sealing. This feature allows resealing to be performed in harmony with normal sealing operations.

7.5.4 Process Qualification

Five HMC packages (2″ × 2″) for qualification of delidding or resealing process for metallic packages were resealed after delidding. Samples were screened as per ISRO-PAS-206 [12]. Delidding of these samples was conducted using delidder equipment. Visual inspection was carried out post delidding process and subsequently the samples were resealed. The above five packages were subjected to evaluation tests and all the samples successfully passed the tests. The details of tests, test conditions, parameters used for delidding and resealing are given in Table 7.8.

Table 7.8 Details of qualification tests

Tests	Test conditions	Result
Sample size: 5 functional HMCs resealed with new lids followed by screening process including 320 h of burn-in at 125 °C		
External visual inspection	As per MIL-STD- 883, Method-2009 [5]	No visual anomalies observed
Electrical test	As per specifications	Meets the specification
Seal leak test	Fine: 30 psia, 24 h. Limit: 5 X10^{-8} atm cc/s Gross30 psia, 24 h. No stream of bubbles	Pass
Temperature cycling	50 cycles −55 °C and 125 °C	Pass
Vibration test (All 3 axes)	Sine vibration: 20 g, 20–100 Hz Sweep rate: 2 oct/min	Pass
	Random vibration 20–2000 Hz Amp: 21.4grms Duration: 2 min/axis	
External visual inspection after vibration test	As per MIL-STD-883, Method-2009	No anomalies observed
Seal leak test	Fine 30 psia, 24 h. Limit: 5 × 10^{-8} atm cc/sec Gross: 30 psia, 24 h. No stream of bubbles	Pass
Group-1		
No. of Samples: 2 (from above 5 samples)		
Liquid thermal shock	5 min–10 s–5 min, 10 cycles	Pass
High temperature bake	1 h at 150 °C	
Moisture resistance test	Method 1004 of MIL-STD-883 (No evidence of pits, discoloration, corrosion)	Pass

7.5.5 Qualification Test Results

Fine leak and Gross leak test result of HMCs before delidding and after resealing of packages were meeting Class 'S' hybrid qualification requirements of MIL-STD-883. The PIND testing carried out and 100% yield is achieved. This validates the process which cannot be validated only by visual inspection.

The main consideration in HMC technology is to have cost effective methodology which necessitates reducing the package cost. The delidding and resealing method is demonstrated which would accommodate most of the hybrids being built for the aerospace applications. Qualification testing has shown that delidding is effective, with no degradation observable after resealing using conventional welding techniques. This paves the way for reusability of the package.

7.6 Bond-Wire in MEMS/Micro-device Packaging

Bond wire is an integral part of IC/MMIC packaging for making connections to other circuitry such as active devices or for input/output connections. Most RF and microwave circuits and sub-systems use bond wires to interconnect components such as lumped elements, planar transmission lines, solid state devices, and ICs [14]. The main aspect of bond wire is the associated inductance which contributes significantly at higher RF frequencies. Bond wire is considered as an extra inductance in circuit level simulation. It is also employed in hybrid MIC to make connections from the innermost section of the printed inductor [15] or connecting active devices. Trimming of MICs is also accomplished by making or breaking the bond wire connection between the components. Bond wire can be single or multiple which is chosen as per the applications. Mostly in MMIC/MIC packaging, bond wires are used to connect input and output connectors so as to reduce iterative and time consuming process of making new mechanical housing. Usage of multiple bond wires provides redundancy in case of breakage of a single wire. Published literatures so far deal with the multiple bond wires keeping into consideration reliability and current carrying capability. Present section modeled the bond wire and provides approaches for finding out the optimum number of bond wires for faithful packaging of micro-devices.

Compared to ribbon, bond wire is having inherent advantage of higher quality factor with the ribbon width equals to wire. Further the selection of the bond pad is of utmost important as the capacitance associated with the pad should not resonate with the inductance of the wire at the frequency of operation resulting in performance degradation. Simple lumped model incorporating the inductance value, series and radiation resistance along with stray capacitance is proposed in the article. Further incorporation of the bond wire in the package and optimum selection for meeting the RF performance is detailed in this article.

7.6.1 Effect of Bond Wire in MEMS Devices

Bond wire introduces primarily inductance at radio frequencies which is to be taken into design consideration. The equivalence of a bond wire is the lumped inductance in series with the resistance. Multiple bond wires are also employed in some applications for reliability criteria but in case of RF it is linked with the reduce inductance and resistance. Also the associated inductance and resistance value of the bond wire varies with the frequency and the effect of the self-inductance and mutual inductance is also significant at higher frequencies. The inductance associated with the bond wire depends on.

- Diameter of wire.
- Material of wire.
- Length of wire.
- Frequency of operation.
- Thickness of wire.
- Height above the substrate.
- Separation between wires.

Mostly, gold material is chosen as it is having higher fusing current compared to copper and aluminum. Further current carrying capability is enhanced with the incorporation of multiple bond wires. In case of multiple bond wires (n), the net inductance is $1/\sqrt{n}$ times of the single wire. The main criteria are the associated radiation at higher frequency which is to be taken into consideration for achieving close analogy between simulated and measured results.

A single wire can be modeled as inductance with series resistance. The radiation associated with higher frequencies is modeled as resistance G shunted with stray capacitance Cs.

Since inductance is determined by the magnetic flux external to wire, the variation in the wire cross section dimensions has little effect on the inductance. Bond wire with smaller cross-section area is having larger inductance due to generation of more magnetic flux. The free space inductance of a wire of diameter d and length ℓ(um) is given as (in the present case there is air beneath)

$$L_e(nH) = 2 \times 10^{-4}\left[\ln\left(\frac{2\ell}{d} + \sqrt{1 + \left(\frac{2\ell}{d}\right)^2}\right) + \frac{d}{2\ell} - \sqrt{1 + \left(\frac{d}{2\ell}\right)^2} + c\right]$$
$$L(\mu H) = 0.0002b\left[\ln\left(\frac{2b}{a}\right) - 0.75\right];\ a(\text{mm}) : \text{wire} - \text{radius};\ b(\text{mm}) : \text{wire} - \text{length}$$
$$c = 0.25\tanh\left(\frac{4\delta}{d}\right);\ \delta = \frac{1}{\sqrt{\pi f \mu \sigma}} \tag{7.2}$$

Considering bond wire at a height of h [13], the given equation is modified as (h, ℓ, d are in mils)

$$L_e(nH) = 5.08 \times 10^{-3}\ell \left\{\ln\frac{4h}{d} + \ln\left[\frac{\ell + \sqrt{\ell^2 + \frac{d^2}{4}}}{\ell + \sqrt{\ell^2 + 4h^2}}\right] + \sqrt{1 + \frac{4h^2}{\ell^2}} - \sqrt{1 + \frac{d^2}{4\ell^2}} - \frac{2h}{\ell} + \frac{d}{2\ell} + c\right\} \tag{7.3}$$

The wire resistance in series can be written as (r is radius, d is the diameter, σ is the conductivity):

$$R_s = \frac{\ell}{2\pi.r\delta\sigma} = \frac{4\ell}{\pi.\sigma.d^2}\left[0.25\frac{d}{\delta} + 0.2564\right] \tag{7.4}$$

$$C_s = \frac{33.33}{60\,\ln\left[\frac{4}{d}\right]}\mathrm{pF/cm} \tag{7.5}$$

$$R_r = 5\pi^2\left(\frac{2\pi a}{\lambda}\right)^4 \tag{7.6}$$

R_r associated at low frequency is in pico ranges which at Ka-band increase to milli ranges value. C_p is associated with capacitance associated between multiple bond wires and generally is in pF range. Effect of higher frequency results in enhanced resistance associated with the bond wire which results in higher losses.

7.6.2 RF Performance Characterization

Wire bonds are used extensively in packaging technology for chip to package or chip to RF connection bonding purpose. At the higher frequency range, it also has some detrimental effect on the RF performance. It can be modeled as a parasitic inductor series with some resistor, if its dimension is shorter than the operating wavelength. When the RF current cycle increases and decreases in a sinusoidal fashion, the magnetic field surrounding a bond wire expands and contracts, inducing a time varying voltage. In bond wire inductance exists both internally (L_{int}) as well as externally (L_{ext}) and with the increase frequency the internal inductance (L_{int}) decreases and external inductance increases due to skin effect ($L_{\mathrm{e}} = L_{\mathrm{int}} + L_{\mathrm{ext}}$). Also the effect of mutual inductance (L_{m}) is to be taken into consideration for multiple bond wires. The mutual inductance is larger for narrower spacing due to the enhanced magnetic coupling whereas diameter of the wire affects series resistance. Wire bonds inductance can be represented as (S is the spacing between wires):

$$L_{\mathrm{T}} = \sum(L_{\mathrm{int}} + L_{\mathrm{ext}}) + \sum L_m \tag{7.7}$$

where

$$L_{\mathrm{m}} = 2\times10^{-4}\ell\left[\ln\left\{\frac{\ell}{S} + \sqrt{1+\left(\frac{\ell}{S}\right)^2}\right\} + \sqrt{1+\left(\frac{\ell}{S}\right)^2} - \sqrt{1+\left(\frac{S}{\ell}\right)^2} + \frac{S}{\ell}\right] \tag{7.8}$$

Ideally, bond wires should be equally spaced to avoid interactive coupling fields affecting the device performance. The skin-depth phenomena also come into picture. So, the dimension of the wire becomes very important. At the frequency above 6 GHz, both the substrate and mutual capacitances play important role. To optimize the number of the bond wires needed between RF transition and Device under Test (DUT), parametric simulation study was performed. Figure 7.8 shows a generalized schematic diagram of the bond wire connected package portion and Table 7.9 summarizes the effect of bond wire numbers on the actual package with a gap of 0.3 mm [17] (Fig. 7.9).

The length of the bond wire also affects the insertion loss and 1.5 dB/mm loss is observed at Ka-band frequency. In the present case $d = 1$ mils, $\ell = 10$ mils and $g = 20$ mils.

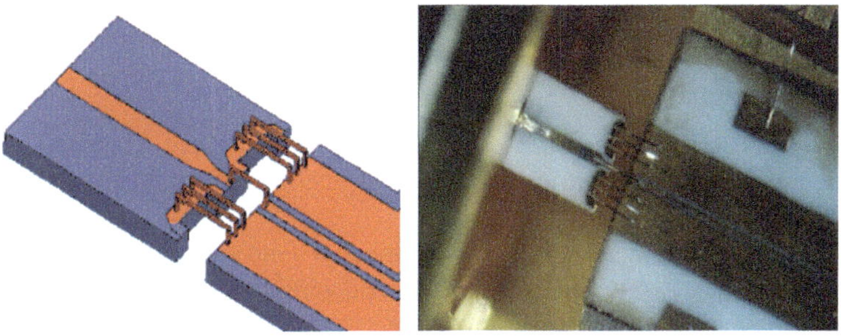

Fig. 7.8 Schematic representation and actual circuit with bond wire

Fig. 7.9 Illustration of bond wire with metallic ground plane insertion

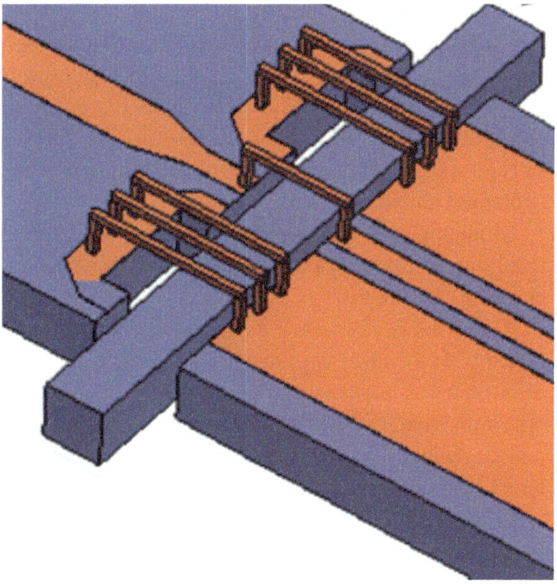

Table 7.9 Losses associated with bond wires at Ka-band

Bond wires	Return loss (dB)	Insertion loss (dB)	Theoretical inductance (nH)
			$(L_e + L_m)$
1	– 8.24	– 3.4	0.350
2	– 12.4	– 2.9	0.247
3	– 18.9	– 2.3	0.202
4	– 11.9	– 2.8	0.175
5	– 18.9	– 2.8	0.156

At higher frequency inductance value decreases as due to skin effect phenomena L_{int} decreases. But the resistance associated with inductance increases due to enhanced surface resistance and radiation. Also it is preferred to have odd number of bond wires (>1) to realize optimum performance as mismatch losses are reduced considerably.

Due to the image current introduced by the ground plane, the overall inductance is reduced due to cancelation of mutual and external inductance. This is collaborated with an experiment by inserting metallic block acting as a ground plane beneath the wire bonds.

As per the above Table 7.9, minimum number of bond wires required is 3 for better RF characteristics. Parallel combination of bond wires reduces the overall inductance which is summation of self and mutual inductance. The overall inductance is shown as

$$L_T = \left(\frac{L_e + L_m}{3}\right) \tag{7.9}$$

To improve further the insertion loss, metallic block acting as a ground plane is inserted which reduces the overall inductance (changes from ~ 0.20 nH to 0.10 nH) and results in the improvement of ~ 1.5 dB insertion losses. This effect is due to reduction of the inductance associated with the image currents phenomena as explained above.

Modeling and characterization of the bond wire using simulation and practical results are explained in this subsection. Simple lumped modeling of bond wire incorporating various aspects is detailed in this part. Bond wire plays an important role in the device performance as at higher frequencies inductance and resistance associated with it causes degraded performances. It was also found out that odd number of bond wires along with reduced gap height results in better RF performance. The optimum match at Ka-band frequency is found out resulting in better impedance match. Authors believe that this study will pave way for more accurate prediction of RF performance at higher RF frequencies using the predicted model and parameter [16, 17].

7.7 Summarization

This chapter details about the various critical aspects of realizing micro-devices. Wafer bonding techniques, especially anodic as well eutectic mechanisms are discussed in detail from critical point of view. Generic process flow in this context is elaborated with standard as well as SOI-based wafers. Various experimental results are briefed here from practical point of view. Further, the packaging aspects in realization of micro or MEMS devices have been chalked out in this chapter. Case studies for high-frequency (RF) MEMS devices have been elaborated. Effect of bond wires in high frequency packaging is elaborated from the essence of performance degradation or optimization.

Questions

1. What are the different wafer bonding mechanisms available for MEMS technology?
2. What are the merits and demerits of 'Anodic Bonding' method?
3. What are the merits and demerits of 'Eutectic Bonding' method?
4. Name some MEMS products where anodic and eutectic bonding methods are implemented.
5. How do you test the quality of the wafer bonding procedure?
6. What precautions should one take prior to do any wafer bonding process?
7. What is 'Delidding' process in packaging method of micro-devices?
8. What are the challenges involved in package delidding process?
9. What is 'Bond-pull' test ? How is it useful to qualify the MEMS process?
10. How is wire bonding plays important role in MEMS packaging?
11. How is wire bonding affects the performance of devices in case of RF or microwave devices?
12. Can the wire bonding be a major cause for giving detrimental effect in RF circuits? How do you mitigate those problems?

References

1. Singh K, Nirmal AV, Sharma SV (2017) Characterization and mitigation of electro-static bonding failures in micro sensors. ICTACT J Microelectronics 3(1)
2. Singh K (2017) Overview of MEMS sensors and associated aspects. ICTACT J Microelectronics 3(2)
3. Singh K (2017) Contaminations in MEMS processes and removal methodology. ICTACT J Microelectronics 3(1)
4. Ko WH, Suminto JT, Yeh GJ (1985) Bonding techniques for microsensors. Micromachining and Micropackaging of Transducers, 1985 Elsevier Science Publishers
5. Singh K, Nirmal AV (2016) Reliability aspects in RF-MEMS circuits for space applications. J Eng Tech Res 4 (6):1–11
6. Petersen K, Barth P, Poydock J (1998) Silicon fusion bonding for pressure sensors. IEEE Solid State Sensor and Actuator Workshop, pp 144–147

7. Hyun SK, Blick RH, Kim DM, Eom CB (2004) Bonding silicon-on-insulator to glass wafers for integrated bio-electronic. Appl Phys Lett 85(12)
8. Knechtel R, Dahlmann G, Schwarz U (2005) Low and High temperature silicon wafer direct bonding for micromechanical absolute pressure sensor. In: Electromechanical Society Proceedings, 2005–06, pp 205–207
9. Licari JJ Hybrid microcircuit technology hand book
10. Harper CA (1974) Hand book of thick film hybrid microcircuits. McGraw Hill Publication, New York, Baltimore Maryland
11. Luce WF Delidding and resealing hybrid microclectronic packages. Manufacturing Methods Report, MIMCOM MM&T
12. Qualification requirement for thick film hybrid microcircuits. ISRO-DOC-No. ISRO-PAS-206Issue-4 NOV-2012
13. Test Method Standards Microcircuits MIL-STD- 883. DOD USA, November 2013
14. Wartenberg SA (2002) RF Measurement of die and packages. Artech House Publication
15. Lim K, Pinel S, Davis M, Sutono A, Lee C-H, Heo D, Obatoynbo A, Laskar J, Tantzeris EM, Tummala R (2002) RF-system-on-package (SOP) for wireless communications. IEEE Microwave Mag 3(1):88–99
16. Bond-wire modeling Standard, EIA/JEDEC Standard No. 59, June 1997, Electronic Industries Association
17. Karmakar A, Singh K (2016) Full-wave analysis and characterization of RF package for MEMS applications. Microw Rev J IEEE MTT-S Serbia Chapter, pp17–22

Annexure A

MEMS Sensors in/with CMOS Technology

MEMS are typically transducer systems that sense or control physical, optical, biological, chemical, or high-frequency quantities, such as acceleration, rotation, inclination, pressure, temperature, relative humidity, radiation, or fluids. To enable the transducer (MEMS) to perform useful functions, the electrical interface with the outside world is, in most of the cases, realized through integrated circuits (ICs). The International Technology Roadmap for semiconductors describes the addition of MEMS functionalities to ICs as an important *'More-Than-Moore'* technology.

MEMS and CMOS can be integrated using two basic methods:

- Multi-chip solutions.
  - 2D side-by-side integration.
  - 3D stacking(system in package solution).
- Single-chip solution or system-on-chip (SoC) solution.
  - Monolithic MEMS and IC integration using MEMS first processing.
  - Monolithic MEMS and IC integration using interleaved MEMS and IC processing.
  - Monolithic MEMS and IC integration using MEMS last processing via bulk micromachining of IC substrate.
  - Monolithic MEMS and IC integration using MEMS last processing via surfacemicromachining.
  - Heterogeneous MEMS and IC integration with via formation during layer transfer.
  - Heterogeneous MEMS and IC integration with via formation after layer transfer (Tables A.1 and A.2).

K. Singh and A. Karmakar, *Micro Electro Mechanical System (MEMS) Technology for Multifaceted Applications*, https://doi.org/10.1007/978-981-96-6162-6

Table A.1 Brief description of CMOS-MEMS integrated solutions for various practical applications

Category	Major product list
Automotive	• Airbags—Accelerometer used for crash sensing • Automatic headlight leveling—Accelerometer used for tilt sensor • Hands-free communication—Silicon microphone • GPS—Gyro sensor • Stability control—Gyro plus accelerometer sensor sense yaw, and compensate for over and under steering on a slick road surface • Tire pressure monitoring systems—Pressure, temperature, and accelerometers assure that tires are properly inflated • Manifold absolute pressure (MAP) sensor—Pressure sensor used for engine control • Seat cushion air bladder—Pressure sensor used for inflating and deflating the bladder • Seat belt pretensioner system—Pressure sensor used to lock the seat belt during a crash
Space sector	• Acoustic sensor for launch vehicle applications • Bending moment accelerometer for launch vehicle application • Accelerometer, gyroscope, magnetometer, etc. for inertial measurement unit(IMU) in space craft • Miniaturized antenna, phase shifter and switch for multifaceted applications • High Q-filters, resonators • Reconfigurable devices, like-tuned filter, resonators, antenna, phased array antenna, etc • On-chip biomedical solutions, i.e., lab-on-chip(LoC) solution • Various sensors for manned missions, like-pressure sensor, temperature sensor, gas sensor, etc • Radiation hardened sensors
Aeronautics/aviation	• Inertial navigation system/GPS—Navigation for unmanned aerial vehicles • Inertial measurement unit(IMU) consisting accelerometer, Gyroscope, magnetometer, Inclinometer, etc • Pressure sensor • Temperature sensor

(continued)

Table A.1 (continued)

Category	Major product list
Consumer	• Sensor detects if device is dropped to prevent hard drive damage(Apple iPods, Toshiba, IBM, and Apple laptops) • MEMS-microphone • Popular gaming console uses MEMS accelerometers in the control device for lifelike gameplay(Nintendo Wii) • Analog devices accelerometer used to monitor terrain for automatic suspension adjustment(Cannondale Simon® Mountain Bike) • Optical MEMS single scanning mirror enables projection capabilities for mobile handsets(Microvision Pico Projector) • MEMS gyroscopes provide image stabilization(Pentax Optio A-30) • Image stabilization(Hitachi DZ-BD70A and DZ-BD7HA camcorders) • MEMS-based motion sensors keep laptops and sensitive data from falling into the wrong hands (Caveo Anti-Theft PC Card) • Computer mouse that can perform in-air functions(Gyration M2000 Travel Air Mouse) • Universal remote that can change a channel with only simple wrist movements(Gyration R4000 LCD Music Remote) • Marine-based night vision system includes optional two-axis MEMS image stabilization • Digital compasses featuring MEMS accelerometers for underwater applications • MEMS gyroscopes sense the passenger's body movements for natural steering(Segway PT Scooter)
Energy	• MEMS inkjet printer that can be used in solar cell production(Dimatix materials printer) • Energy harvesters • SmartHome control systems

(continued)

Table A.1 (continued)

Category	Major product list
Industrial applications/environmental sensing	• MEMS print heads for printing billboards and other large signage(HP Scitex XL2200 Industrial Wide-format Printer) • Optical device that draws patterns on wafers in the mass semiconductor production process • Portable near-infrared analysis (NIR) tool uses a MEMS-based optical modulator for chemical identification MEMS strain gauge pressure sensor—Gauges pressure on hydraulic lines in harsh environment construction and earth mover equipment • 3-axis Accelerometer
Medical	• Shunt valve regulation—Magnetic switch used to adjust the shunt valve without further surgery • Miniaturized insulin delivery system • Blood pressure monitor (invasive and non-invasive)—Pressure sensor used to sense pulse rate • MEMS accelerometers measure patient's arm angle for more accurate reading • Wireless AAA pressure sensor—Monitors pressure on aortic wall to aid in aneurysm repair • Inhalers—Pressure sensor used to control drug dispensing • Micronozzle-based instrument for creation of small size distribution of droplets • Sleep apnea mask—Pressure sensor used to measure flow of air (breathing) during sleep • Kidney dialysis machines—Pressure sensor used for pumps • Catheters—Pressure sensor used for pressure and flow measurement • Adjustable hospital beds—Pressure sensor used in the mattress air bladder • Adaptive rate pacemaker—Accelerometer used to sense exertion and regulate heart rate accordingly • MEMS camera/antenna module for wireless capsule endoscopy • Timed drug delivery – MEMS implantable chips • Hearing aid—MEMS-microphone and speaker • Lithium detectors—Checking lithium content in blood for polar disorder diseases • Micro-mirror: Adaptive optics systems for medical and space applications

Table A.2 Abridged information about various foundries round the globe dealing with CMOS-MEMS activities

Details of company	Substrate material used	Wafer diameter (inch)	Situated in country
Asia pacific microsystems, Inc	Silicon, Sapphire, SOI, Glass, Quartz, Polymer, CMOS-wafers	6	Taiwan
Atomica	Silicon, SOI, borosilicate, fused silica, quartz, glass, piezo, and engineered substrates	6, 8	United States
C2MI	Silicon, SOI, glass, and CMOS-wafers	8	Canada
CEA-Leti	Silicon, SOI, glass, and CMOS-wafers	8, 12	France
CSEM	Silicon, SOI, lithium niobate, silicon carbide, fused silica, quartz, glass	6	Switzerland
Fraunhofer institute for silicon technology (ISIT)	Silicon, SOI, glass, GaN and CMOS	8	Germany
Fraunhofer institute for microelectronic circuits and systems (IMS)	Silicon, SOI	8	Germany
Hanking electronics	Silicon	8	China
IMEC	Si, SOI, fused silica or glass; post-processing on CMOS-wafers (8-inch to 12-inch)	8, 12	Belgium
Micronit	Silicon and SOI, borosilicate glass or Polymer; structured wafer	6, 8	Netherlands
LioniX international	Silicon, SOI, fused silica, borosilicate glasses, thermally oxidized silicon	4, 6	Netherlands
Mapper	Silicon, SOI and various glasses, piezoelectric materials	4	Russia
Microfluidic foundry	Silicon, Glass, SU8/ PDMS, Polymer		United States
nanoPHAB	III-V materials, GaAs, InP, Si, Oxi-Nitrides, Glass, Polymers, Multimaterials	2, 3, 4	Netherlands

(continued)

Table A.2 (continued)

Details of company	Substrate material used	Wafer diameter (inch)	Situated in country
Nanoshift LLC	Silicon, SOI, Glass, flex substrates, polymer, and other materials	2, 3, 4, and 6	United States
NEMS and MEMS Proto Fab (NMPF)	Silicon, Glass, Quartz, SOI wafers	2, 4	India
Philips	Silicon, SOI, glass, quartz and CMOS-wafers	6, 8	Netherlands
Robert bosch GmbH	Silicon, SOI, Fused silica or glass, post-processing on 200 mm-CMOS-wafers	8	Germany
Safran sensing technologies	Silicon, SOI	6	Switzerland, Norway
Science foundry	Silicon, SOI, glass, quartz, polymer	6	United States
Semefab		4, 6	UK
Semiconductor laboratory (SCL) under MeitY/GoI	Silicon, SOI and various glass(Pyrex, Borosilicate)	6,8	India
Silex microsystems	Silicon	8	Sweden, China
Silicon sensing systems	Silicon	8	UK, Japan
SINTEF MiNaLab	Silicon, SOI, Glass	6	Norway
Sony semiconductor	Silicon, SOI	6, 8	Japan
STMicroelectronics	Silicon		Switzerland
Teledyne DALSA	Silicon and polysilicon	6, 8	Canada
Teledyne Micralyne	Silicon, SOI and various glasses	6	Canada
Tower semiconductor	Silicon	6, 8	Israel
TDK Tronics	Silicon, SOI and glass wafer	6	France, Japan
TSMC	Silicon	6, 8, 12	Taiwan
uFluidix	Silicon, glass, PDMS		Canada
United microelectronics Corporation	Silicon	8	Taiwan
Wunderlichips	Silicon, SU8\PDMS	4, 6, 8	Switzerland
X-Fab	Silicon and glass	6, 8	Germany

References

1. Fischer A, Forsberg F, Lapisa M et al (2015) Integrating MEMS and ICs. Microsyst Nanoeng 1:15,005. https://doi.org/10.1038/micronano.2015.5
2. https://en.wikipedia.org/wiki/List_of_MEMS_foundries

Annexure B

MEMS Market Size

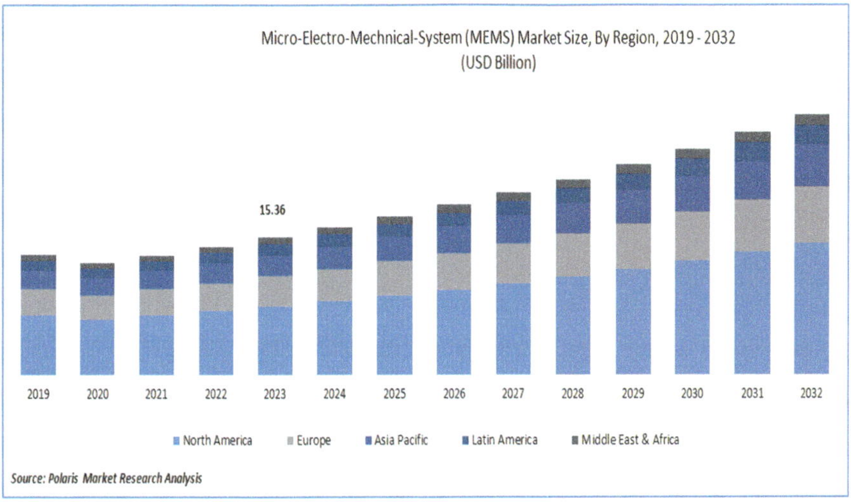

Top Companies Market Share in MEMS Sensor Industry: (In no particular order of rank).

- STMicroelectronics NV
- InvenSense Inc. (TDK Corp.)
- Bosch Sensortec GmbH (Robert Bosch GmbH)
- Analog Devices Inc.
- Murata Manufacturing Co. Ltd
- Kionix Inc. (ROHM Co Ltd)

K. Singh and A. Karmakar, *Micro Electro Mechanical System (MEMS) Technology for Multifaceted Applications*, https://doi.org/10.1007/978-981-96-6162-6

- Infineon Technologies AG
- Freescale Semiconductors Ltd (NXP Semiconductors NV)
- Panasonic Corporation
- Omron Corporation
- First Sensor AG (TE Connectivity)

Market Size for various MEMS-based Sensors for the period of 2019–2023[1, 2].

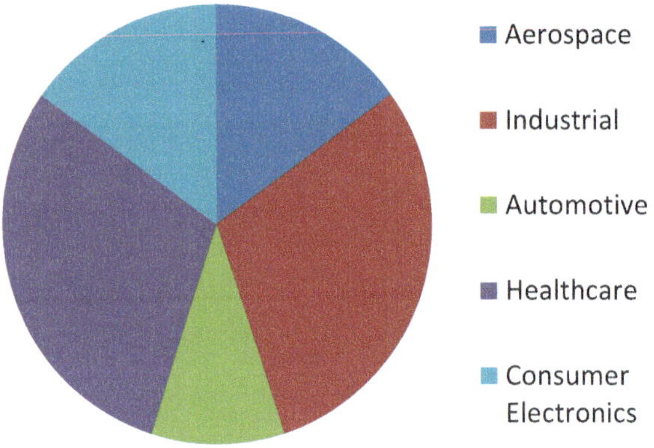

References

1. https://www.cognitivemarketresearch.com/mems-sensor-market-report
2. https://www.mordorintelligence.com/industry-reports/mems-market/news

Annexure C

Standard R&QA Tests for MEMS Sensors

One of the challenges of MEMS technology is its reliability. To ensure the end usage of the sensors in the prescribed environments, various accelerated tests are performed on the devices during its qualification stages. Only after successful completion of qualification test, screened sensors shall be cleared for use in end usage of subsystem. Here, glimpse of screening and qualifications tests have been jotted down.

1. Screening Plan:

All devices of both types (Types 1 and 2) in procured batch shall be subjected to screening test to eliminate potential weak devices. Table C.1 gives the screening test plan.

2. Qualification Test Plan:

Sensors from screened batch shall be subjected to lot qualification test as per plan given in Table C.2.

Vibration and Mechanical Shock Test Requirement

Sine Vibration

The device(s) shall be rigidly fastened on the vibration platform and the leads adequately secured. The devices(s) shall then be subjected to vibration test as per following conditions.

Method: MIL-STD-883, Method-2007, Condition-A.
Acceleration level: 20 g.

K. Singh and A. Karmakar, *Micro Electro Mechanical System (MEMS) Technology for Multifaceted Applications*, https://doi.org/10.1007/978-981-96-6162-6

Table C.1 Screening test plan

SR.#	Test	Method/condition
1	External visual inspection	@10x
2	Radiographic inspection	MIL-STD-883 Method-2012
3	Functional characterization	As per datasheet
4	Stabilization bake	MIL-STD-883 Method-1008, At maximum storage temperature of device
5	Thermal cycling	MIL-STD-883 Method-1010, 10 cycles between min and max storage temperature of device
6	Operational Thermo-vac cycling	5 cycles between min and max operating temperature of device
7	Random vibration	As per Section-A.1
8	External visual inspection	@10x
9	Pre-burn-in functional characterization	As per datasheet
10	Burn-in	At rated operating conditions
11	Post-burn-in functional characterization	As per datasheet
12	External visual inspection	@10x
13	Radiographic inspection	MIL-STD-883 Method-2012

Cycle: 20 Hz to 2000 Hz, then back to 20 Hz.
Cycle number: 4 cycles per axis.
Cycle duration: 4 min.
Axis number: 3
Bias condition: With and without bias.

Random Vibration

Method: MIL-STD-883, Method-2026.
gRMS Value:9.8gRMS for screening.
13.0gRMS for qualification.
Duration: 2 min for screening.
15 min for qualification.
Axis number: 3
Bias condition: With and without bias.

Mechanical Shock

Method:MIL-STD-883, Method-2002, Cond. B
Level:1500 g.
Pulse duration:0.5 ms.

Table C.2 Qualification test plan

Sn	Test	Method/condition	Accept (reject) criteria
A	**Functional Characterization**		**33 Devices**
	Functional characterization	As per datasheet	33(0)
B	**Destructive Physical Analysis (DPA)**		**3 Devices**
	External visual inspection	@10x	3(0)
	Physical dimension	As per datasheet	3(0)
	Radiographic inspection	MIL-STD-883 Method-2012	3(0)
	Resistance to solvents	MIL-STD-883 Method-2015	3(0)
	Connector mating/demating test	Condition decided by end-users	3(0)
	Decapsulation	As per end-user procedure	3(0)
	Internal visual inspection for DPA	MIL-STD-883 Method-2013	3(0)
	Destructive bond pull test	MIL-STD-883 Method-2011	22(0) or all bond wires whichever is less on 2 devices
	Die shear strength or Substrate attach strength	MIL-STD-883 Method-2019 Or Method-2027	3(0)
	Lead integrity or Plated through hole (PTH) quality assessment (as applicable)	MIL-STD-883C Method-2004 Cond. B2	3(0)
	Solderability	MIL-STD-883 Method-2003	3(0)
C	**Thermal Group**		**3 Devices**
	External visual inspection	@10x	3(0)
	Radiographic inspection	MIL-STD-883 Method-2012	3(0)
	Thermal cycling	MIL-STD-883 Method-1010, 100 cycles between min and max storage temperature	3(0)
	Active thermo-vacuum cycle (with gas analysis using mass spectrometer)	25 cycles between min and max operating temperature	3(0)
	Moisture resistance test	MIL-STD-883 Method-1004	3(0)
	External visual inspection	@10x	3(0)
	Radiographic inspection	MIL-STD-883 Method-2012	3(0)
	Outgassing test	ASTM-E-595	1(0)
D	**Mechanical group**		**3 Devices**
	External visual inspection	@10x	3(0)
	Radiographic inspection	MIL-STD-883 Method-2012	3(0)

(continued)

Table C.2 (continued)

Sn	Test	Method/condition	Accept (reject) criteria
	Sine vibration test	As per Section-A.1	3(0)
	Random vibration test	As per Section-A.1	3(0)
	Mechanical shock test	As per Section-A.1	3(0)
	External visual inspection	@10x	3(0)
	Radiographic inspection	MIL-STD-883 Method-2012	3(0)
E	**Endurance life group**		**15 Devices**
	External visual inspection	@10x	15(0)
	Radiographic inspection	MIL-STD-883 Method-2012	15(0)
	Life test	1000 Hrs Operation / 1000 Operation Cycles @ maximum operating temperature	15(0)
	External visual inspection	@10x	15(0)
	Radiographic inspection	MIL-STD-883 Method-2012	15(0)
F	**Radiation hardness assurance group**		**9 Devices**
	External visual inspection	@10x	9(0)
	Radiographic inspection	MIL-STD-883 Method-2012	9(0)
	TID	127 rad**	**3(0)**
	Proton for displacement damage	10 MeV, 7.56E + 07 p/cm^2**	**3(0)**
	Heavy ion	LET of 80 MeV-cm^2/mg**	**3(0)**
	External visual inspection	@10x	9(0)
	Radiographic inspection	MIL-STD-883 Method-2012	9(0)
G	**Tests for human rating**		3 Devices
	Flammability	International standards like: NASA-STD-6001B	1(0)*
	Toxicity		1(0)*
	Odor		1(0)*

*One device each from Group-C, D, and E shall be subjected to these tests

Shock number:5 / direction / axis.
Direction number:2 / axis.
Axis number:3
Total shock number: 30.

Bias Condition: With and without bias.

Annexure D

MEMS in Space

MEMS technology is driving the miniaturization of various mechanical or movable parts and the development of full system-on-chip (SoC) devices. With all promising features, this technology is becoming widely popular in multifaceted applications for space usage in recent era. However, the radiation in space may affect significantly with these portable devices by altering its inherent electrical/mechanical properties. The susceptibility of MEMS devices with space radiation is depicted in Table D.1, whereas Table D.2 describes the level of dose in various satellite orbits.

References

1. Hartzell A, da Silva M, Shea H (2011) [MEMS Reliability], Springer. ISBN 978-1-4419-6017-7
2. Iniewski K (2010) [Radiation Effects in Semiconductors], CRC Press
3. Shea H (2009) "Radiation sensitivity of microelectromechanical system devices", J. Micro/Nanolith. MEMS MOEMS 8(3), 031303

K. Singh and A. Karmakar, *Micro Electro Mechanical System (MEMS) Technology for Multifaceted Applications*, https://doi.org/10.1007/978-981-96-6162-6

Table D.1 Effects of space radiation on various integrated MEMS devices

MEMS device	Actuation type	Minimum dose for failure	Radiation type	Failure mode
Analog Devices ADXL 50	Electrostatic (comb-drive)	25 krad (Si)	Co-60 γ	Dielectric charging in device
Analog Devices ADXL 50	Electrostatic (comb-drive)	> 50 krad (Si)	SEM localized e-beam 30 keV	Dielectric charging in device
Analog Devices ADXL 50	Electrostatic (comb-drive)	100 krad (Si)	5.5 MeV protons	Dielectric charging in device
Analog Devices ADXL 50	Electrostatic (comb-drive)	20 krad (Si)	65 MeV protons and heavy ion	Dielectric charging in device
Endevco accelerometer 7264B-500 T	Piezo-resistive	> 30 Mrad	Co-60 γ	Trapped charge, depletion of minority carriers
Kulite pressure transducers XTE-190-25A	Piezo-resistive	7 Mrad to > 20 Mrad, sample dependent	Co-60 γ	Trapped charge, depletion of minority carriers
Rockwell Scientific Co RF-Switch	Electrostatic (parallel-plate)	30 krad (GaAs)	Co-60 γ	Dielectric charging in device (strongly geometry dependent)
FBK-IRST ohmic RF-Switch	Electrostatic (parallel-plate)	10 Mrad (SiO_2) proton 1 Mrad (SiO_2) X-ray	2 MeV protons 10 keV X-ray	Both NIEL and ionizing damage
NASA /GSFC Microshutter array	Electrostatic and electromagnetic	10 to > 200 krad (Si) depending on drive voltage	Co-60 γ and 60 K	Charge trapping (dielectric charging)

Table D.2 Annual radiation doses for different satellite Orbits

Type of orbit	Annual dose deposition	Predominant particles
Lower earth orbit (LEO)	> 100 krad	Trapped electrons
Geostationary orbit (GEO)	> 10, 000 krad	Trapped electrons

Annexure E

Thin-Film Materials and Gases

See Tables E.1 and E.2.

Table E.1 Thin-film materials deposited by evaporation technique used in MEMS/NEMS

Material	Minimal evaporation temperature	State of evaporation	Typical deposition rate(nm/s)
Aluminum	1010	Melts	2
Al_2O_3	1325	Semimelt	1
Carbon	2140	Sublimes	3
Chromium	1157	Sublimes	1.5
Copper	1017	Melt	5
Gold	1132	Melt	3
Platinum	1747	Melt	2
TiN	997	Melt	1
Titanium	1453	Melt	2
Nickel	1262	Melt	2.5
Indium	742	Melt	10
Iron	1180	Melt	5

K. Singh and A. Karmakar, *Micro Electro Mechanical System (MEMS) Technology for Multifaceted Applications*, https://doi.org/10.1007/978-981-96-6162-6

Table E.2 Safety-related attributes of precursor gases

Gas	Corrosive	Flammable	Pyrophoric	Toxic	Bodily hazard
Ammonia (NH_3)	×			×	Eye and respiratory irritation
Boron tricholoride (BCl_3)	×				
Chlorine (Cl_2)	×			×	Eye and respiratory irritation
Hydrogen fluoride (HF)	×				Severe burns
Hydrogen (H_2)		×			
Silane (SiH_4)		×	×	×	
Silicon tetracholoride ($SiCl_4$)	×				

Annexure F

Water in Semiconductor Industry

In the field of microelectronics or semiconductor, precision and reliability are paramount. This industry is exponentially expanding with technological advancements in terms of smaller, faster, and less power hungry devices. To ensure flawless operation of these delicate devices, the quality or grade of materials used is of utmost importance. One such essential resource is water, especially purified water treated as 'life-line' for semiconductor industry. In these industries, clean room facilities are the epicenter of production. The clean room must maintain extremely low levels of airborne contaminants, as even microscopic particles can disrupt the fabrication of electronic components. Purified water plays a pivotal role in these clean rooms, serving following critical functions:

- Wafer cleaning.
- Chemical mixing and dilution.
- Cooling and heat transfer.

There are basically two grades of semiconductor water:

- Ultra-pure water (UPW). It is used for most of the semiconductor wafer processing, cleaning, chemical dilution, mixing, etc. The resistivity of UPW is generally 18.2 MΩ-cm at standard room temperature.
- Highly pure water (HPW). It is generally used for cooling and general purpose equipment cleaning.

Several water purifications techniques are adopted in semiconductor industries, which are as follows:

- Reverse osmosis (RO).
- Deionization (DI).
- Ultra-filtration (UF).

K. Singh and A. Karmakar, *Micro Electro Mechanical System (MEMS) Technology for Multifaceted Applications*, https://doi.org/10.1007/978-981-96-6162-6

- Distillation.
- Electro-deionization (EDI).

Because of huge water consumption in semiconductor industry, sometimes it is known as '*A thirsty business.*' Statistical data of water consumption in semiconductor fabline is given below:

To create an integrated circuit(IC) on a 300 mm wafer, can require approximately 2200 gallon of water, including 1500 gallon of UPW. So, a large fab facility that process say, 40,000 wafers a month, can use up to 4.8 million gallons of water per day, this equates to the annual water consumption of a city of 60,000 people.

References

1. https://chinawaterrisk.org/resources/analysis-reviews/8-things-you-should-know-about-water-and-semiconductors/

2. https://www.semiconductor-digest.com/water-supply-challenges-for-the-semiconductor-industry/

3. https://purite.com/the-critical-role-of-water-in-semiconductor-and-microelectronics-production/

Annexure G

Conductivities of Some Materials

Material	Conductivity S/m (20 °C)
Aluminum	3.816×10^7
Brass	2.564×10^7
Bronze	1.00×10^7
Chromium	3.846×10^7
Copper	5.813×10^7
Distilled water	2×10^{-4}
Germanium	2.2×10^6
Gold	4.098×10^7
Graphite	7.0×10^4
Iron	1.03×10^7
Mercury	1.04×10^6
Lead	4.56×10^6
Nichrome	1.00×10^6
Nickel	1.449×10^7
Platinum	9.52×10^6
Silver	6.173×10^7
Steel (stainless)	1.1×10^6
Solder	7.00×10^6
Tungsten	1.825×10^7
Zinc	1.67×10^7

K. Singh and A. Karmakar, *Micro Electro Mechanical System (MEMS) Technology for Multifaceted Applications*, https://doi.org/10.1007/978-981-96-6162-6

Annexure H

Dielectric Constants and Loss Tangents of Some Materials

Material	Frequency (GHz)	ϵ_r	tanδ (25°C)
Alumina (99.5%)	10	9.5–10	0.0003
Ceramic (A-35)	3	5.60	0.0041
Fused quartz	10	3.78	0.0001
Gallium Arsenide	10	12.8	0.006
Glass (Pyrex)	3	4.82	0.0054
Nylon (610)	3	2.84	0.012
Polythene	10	2.25	0.0004
Silicon	10	11.8	0.001
Teflon	10	2.08	0.0004
Vaseline	10	2.16	0.001
Water (distilled)	3	76.7	0.157

K. Singh and A. Karmakar, *Micro Electro Mechanical System (MEMS) Technology for Multifaceted Applications*, https://doi.org/10.1007/978-981-96-6162-6

Annexure I

Thermal Conductivity of Some Materials

Material	Thermal conductivity (Watt/cm × °C)
Silver	4.08
Copper	3.94
Gold	2.96
Aluminum	2.18
Tungsten	1.74
Brass	1.11
Chromium	0.937
Nickel	0.920
Platinum	0.716
Titanium	0.219
RT/Duroid-5880	0.0026
G10/FR4	0.0027
Aluminum nitride	1.70
Sapphire	0.46
Aluminum oxide 99.6%	0.36
Aluminum oxide 96%	0.26
Aluminum oxide 91%	0.13
Glass	0.015
Mica	0.0043–0.0062
Air	0.00026

(continued)

K. Singh and A. Karmakar, *Micro Electro Mechanical System (MEMS) Technology for Multifaceted Applications*, https://doi.org/10.1007/978-981-96-6162-6

(continued)

Material	Thermal conductivity (Watt/cm × °C)
Tin lead solder	0.4921
Silver filled epoxy	0.0156
Epoxy	0.0099
Thermal grease	0.0042–0.0049

GPSR Compliance

The European Union's (EU) General Product Safety Regulation (GPSR) is a set of rules that requires consumer products to be safe and our obligations to ensure this.

If you have any concerns about our products, you can contact us on ProductSafety@springernature.com

In case Publisher is established outside the EU, the EU authorized representative is:

Springer Nature Customer Service Center GmbH
Europaplatz 3
69115 Heidelberg, Germany

Batch number: 08999295

Printed by Printforce, the Netherlands